Praise for *Cold-Hardy Fruits and Nuts*

"Outrageous diversity throughout the ecosystem will always be essential for growing healthy fruit and nuts. Allyson Levy and Scott Serrano have provided a thorough guide to adding an intriguing selection of productive plantings to your landscape. Get on board and plant more of everything!"

—**Michael Phillips**, author of *The Holistic Orchard* and *Mycorrhizal Planet*

"At a time when we are rediscovering the value of biodiversity and habitat enrichment, *Cold-Hardy Fruits and Nuts* offers a compendium of lesser-known backyard forageables that can turn a home garden into a homestead landscape ripe with flavor and nutrition."

—**John Forti**, author of *The Heirloom Gardener*;
executive director, Bedrock Gardens

"Allyson and Scott are deeply committed plant lovers, and this well-researched, handsome book—with educational photographic portraits of each plant—shares their botanical love affair with us! Hortus Arboretum and Botanical Gardens is their magical plant playground that we get to visit, and this delicious gift is birthed from that rich source. They deftly guide us to diversify our landscape with beauty we can eat, while increasing our personal and ecological health."

—**Dina Falconi**, author of *Foraging and Feasting* and
Earthly Bodies and Heavenly Hair

"This is *the* book for growers looking to expand their range of perennial and tree crops. Levy and Serrano have not wasted too much space on general cultivation tips. Instead, they wisely dedicate the bulk of the book to detailed, practical advice on a wide variety of species, some of which I had never come across before, others which I had assumed would need warmer conditions. You will be inspired!"

—**Ben Raskin**, author of *The Woodchip Handbook*;
head of horticulture and agroforestry, the Soil Association

"Allyson Levy and Scott Serrano are consummate gardeners and artists. They bring horticultural prowess and designer insights to this beautiful, easy-to-use, and easy-to-read book. Their in-depth research and hands-on knowledge are invaluable to those looking to expand their palette of plants. Featuring taste profiles, landscape tips, and propagation techniques, *Cold-Hardy Fruits and Nuts* is my new go-to guide for growing these plants. It is wonderful to see native trees and shrubs included with such enthusiasm and appreciation."

—**Marc Wolf**, executive director, Mou

"This book is filled with enthusiasm for growing uncommon fruits and nuts and I can't agree more. Allyson and Scott give honest assessments of each plant based on years of hands-on experience. Along with the detailed plant and fruit descriptions, I particularly like their 'Growth Difficulty Rating' and 'Taste Profile and Uses' sections to help readers decide what to grow. They have me dreaming of growing Himalayan chocolate berries and Korean stone pines."

—**Charlie Nardozzi**, author of *Foodscaping*

"*Cold-Hardy Fruits and Nuts* offers practical experience and useful information on a great diversity of species, including a few surprises. A great place to turn for anyone developing a perennial edible landscape, one of the world's highest-carbon forms of gardening and farming."

—**Eric Toensmeier**, coauthor of *Edible Forest Gardens*, author of *Perennial Vegetables*

"Allyson Levy and Scott Serrano have been my 'go-to' source for information on unusual or common plant subjects to draw. Now they have made their wealth of knowledge available in this beautiful and concise book. Full of essential information alongside interesting facts about each tree, shrub, or vine, the book tells us what to plant and why, how to best grow it, when to harvest, how to crack the nuts, and what to do with an abundance of fruit to preserve. The photos are lovely and descriptive. As an artist, a harvester, and a novice gardener, I now have all the information I need about my fruit and nut growing in one book!"

—**Wendy Hollender**, botanical artist and author of *The Joy of Botanical Drawing*

"*Cold-Hardy Fruits and Nuts* covers a full range of edible plants for the home landscape—from the familiar to the lesser known to the truly exotic. Highly recommended for its great level of detail. The authors' deep research into botanical history and descriptions both satisfies my curiosity and makes me want to delve even deeper into the information they provide."

—**Robert Kourik**, author of *Sustainable Food Gardens* and *Designing and Maintaining Your Edible Landscape Naturally*

Cold-Hardy Fruits and Nuts

Cold-Hardy Fruits and Nuts

50 Easy-to-Grow Plants for the Organic Home Garden or Landscape

ALLYSON LEVY AND
SCOTT SERRANO

Chelsea Green Publishing
White River Junction, Vermont
London, UK

Commissioning Editor: Michael Metivier
Project Manager: Patricia Stone
Project Editor: Benjamin Watson
Copy Editor: Laura Jorstad
Proofreader: Angela Boyle
Indexer: Shana Milkie
Designer: Melissa Jacobson

Printed in the United States of America.
First printing February 2022.
10 9 8 7 6 5 4 3 2 1 22 23 24 25 26

Our Commitment to Green Publishing

Chelsea Green sees publishing as a tool for cultural change and ecological stewardship. We strive to align our book manufacturing practices with our editorial mission and to reduce the impact of our business enterprise in the environment. We print our books and catalogs on chlorine-free recycled paper, using vegetable-based inks whenever possible. This book may cost slightly more because it was printed on paper that contains recycled fiber, and we hope you'll agree that it's worth it. *Cold-Hardy Fruits and Nuts* was printed on paper supplied by Versa that is made of recycled materials and other controlled sources.

Library of Congress Cataloging-in-Publication Data

Names: Levy, Allyson, author. | Serrano, Scott, author.
Title: Cold-hardy fruits and nuts : 50 easy-to-grow plants for the organic home garden or landscape / Allyson Levy, Scott Serrano.
Other titles: 50 easy-to-grow plants for the organic home garden or landscape
Description: White River Junction, Vermont : Chelsea Green Publishing, 2022. | Includes bibliographical references and index.
Identifiers: LCCN 2021051824 (print) | LCCN 2021051825 (ebook) | ISBN 9781645020455 (paperback) | ISBN 9781645020462 (ebook)
Subjects: LCSH: Organic gardening—Handbooks, manuals, etc. | Fruit-culture—Handbooks, manuals, etc. | Nut trees—Handbooks, manuals, etc. | Fruit trees—Handbooks, manuals, etc. | Handbooks and manuals.
Classification: LCC SB357.24 .L48 2022 (print) | LCC SB357.24 (ebook) | DDC 635/.0484—dc23/eng/20211116
LC record available at https://lccn.loc.gov/2021051824
LC ebook record available at https://lccn.loc.gov/2021051825

Chelsea Green Publishing
85 North Main Street, Suite 120
White River Junction, Vermont USA

Somerset House
London, UK

www.chelseagreen.com

To Zoë and Emmett

Who let us buy plants instead of new clothes and toys,
put up with late dinners, and living in a house full of plants.

To Mom, our biggest fan, whose enthusiasm and support
are as rich as the best aged leaf mulch.

Contents

Introduction

We created our botanical garden and arboretum by accident in 1999, after moving to Stone Ridge, a rural community two hours north of New York City. Because we are both visual artists, we began surrounding our home with different plants selected as inspiration for our artwork. The natural world became the main focus of our art as a source of creativity and also as a material through the direct use of dried plant seeds and leaves in our artworks.

Although our methods of making botanically oriented art are completely different, our two paths converged as we increasingly spent more time in the garden and less time on our art. We started to plant native trees, shrubs, perennials, and edible plants; we ran deer fencing around 3 acres (1.2 ha) of land on our property. When the number of new species exponentially increased beyond our ability to remember the individual Latin names of them all, we created detailed metal plant tags and attached them to our trees. To keep track of our collection, we started a comprehensive plant list and began to think about our garden as separate series of plant collections.

It didn't take long for us to discover that the selection of both ornamental and edible plants available for sale in most garden centers and nurseries was reduced to a handful of tried-and-true species that were considered hard to kill. At this point we became interested in the wider range of food plants that would reflect the diversity of edible plants on Earth, and this has remained one of the fundamental goals of our garden. Along the way we discovered that many of the beautiful, decorative flowering plants were also edible. This led us to the joy of researching obscure specialty mail-order nursery catalogs for different edible plants that would thrive in our zone 6 environment, such as pawpaw, arctic kiwi, medlar, and American persimmon. The exorbitant cost of ordering large amounts of rare mail-order plants led us to start a garden landscaping company in order to cover what we were spending on our new obsession. Along the way we also developed an interest in learning how to use our property as a working ecological system for food, and started to inoculate shiitake logs, keep chickens for eggs, and regularly tap the wonderful native maple trees on our land for our family's maple syrup.

In 2009 we purchased the 8 wooded acres (3.2 ha) of land across the road from our house and crowded gardens, with the intent of creating even more crowded gardens! Food plants that had been on our wish list for years, such as blight-resistant American chestnuts, schisandra vine, Korean stone pine, and Asian pears, could now be planted since we had space for them. It was a large, arduous project to clear 3 more acres (1.2 ha) of brush and tree debris, but it provided us with the opportunity to thoughtfully design and create a series of specific plant collections. Over time these included nut trees, a beach plum hedge, and a large Chinese edible plant collection. More important, we started to collect and plant endangered and threatened species.

We now think of our gardens as a type of edible experimental station, featuring marginal food species that were not believed to be hardy to our region, including some of the best surprises: toon (specifically, *Toona sinensis*), Szechuan peppercorn, Himalayan chocolate berry, and maypop (a.k.a. American passionfruit), all of which have produced food for us.

Because of our interests in self-sustainability, plant diversity, and regenerative gardening, we began sharing our gardens with the public via open garden days, garden tours, and fruit growing classes, and have hosted PowerPoint talks at garden clubs and arboretums in our region.

In 2017 the gardens received Level II arboretum accreditation by the ArbNet Arboretum Accreditation Program established by the Morton Arboretum. We now consider ourselves a young arboretum / botanical garden, with the goal of creating a "Living Textbook" of the diversity of plant life that can be grown in the Northeast. In 2019 our gardens became a nonprofit organization and were formally renamed the Hortus Arboretum and Botanical Gardens.

Why Write This Book?

This book is a logical extension of our botanical garden's mission as an educational institution, and will allow us to reach people who cannot attend our classes or visit our gardens. We hope that this publication on fruiting plants will allow us to share our planting experiences with a broader group of gardeners.

Readers may notice that we have not included many of the standard garden favorites (such as apples), nor most of the trees in the

Prunus genus (such as peaches, plums, apricots, and cherries) in this book. This is for four reasons. First, there are already many excellent books devoted to trees like apples, and there is no need to repeat all of the great literature on well-known fruit like peaches. You can also find more reference material on fruiting and food plants in appendix 1, located in the back of this book.

Second, and more important, as part of creating a sustainable environment, landscaping your garden or home environment with a greater diversity of plant life will not only provide you with more food but also will provide pollinating insects with a greater diversity of flower pollen choices over an entire growing season. Nature and gardeners alike benefit from more diversity.

Third, we believe that because of climate change, a more diverse food palette will not only provide gardeners with new and interesting fruit choices, but also give the human race more adaptability and planting options as our climate begins to heat up and change in unexpected ways.

Fourth, based on our hands-on experience with growing fruits such as peaches, many people find that after planting a peach tree, they are consistently plagued by a large spectrum of diseases and insect pests. In our neck of the woods, wild cherry trees are a common and important part of the forest ecosystem, which means that there are literally hundreds of insects that like to eat their foliage and dozens of pathogens that use them as host plants. While wild cherry trees are a crucial food source for wildlife, they mean bad news for people who garden around them. Any pest that attacks a wild or cultivated cherry tree near your garden will eventually attack peaches, apricots, plums, and cherries, which are closely related to that species. While we do grow a few peach trees, we understand that in order to get healthy fruit from them, we have to constantly watch and care for them to keep them healthy. We feel that these types of plants will discourage many gardeners because they become riddled with pests.

Why not try planting a few food plants that are less likely to be a host for hundreds of pests? This is why we are advocates for planting all of the other wonderful fruiting edible plants that are basically pest-resistant. We grow each of the fifty plants represented in this publication, and the unifying element of all of them is that they are all about 90 to 95 percent *pest-free*. Although every plant attracts some pests, these plants are not as susceptible to the problems that plague trees like peaches and apples. So by all means plant a peach tree if you love peaches, but also try to make room in your garden for some of these other wonderful alternative fruiting plants, because they will provide you with a broad range of food choices that require little maintenance.

Our hope is that readers will also "see" wonderful discoveries among these various fruiting plants. We have devoted a lot of the space in this book to the images of the actual plants. Many of the books on edible plants have only small images that do not do justice to the unique beauty of each individual species. Because of our backgrounds as visual artists, we hoped to highlight the aesthetic aspects of each plant as well as how it tastes. Hopefully, readers will discover that many of these fifty plants will fill their gardens with beautiful flowers that are worth including around their homes.

Most of all, we hope that you will use this book as a source of inspiration and a helpful guide for creating your own personal edible environment.

How to Use this Book

This book is the result of our years of firsthand experience growing fruiting plants, sometimes killing a plant several times to figure out exactly what it requires to survive in our environment. Although we focus on lower-maintenance edible plants, we feel that all plants will need some amount of attention in the first year or two to enable them to flourish. The "Growth Difficulty Rating" in each chapter is a guide to understanding both the cultural needs of the plant as well as the level of maintenance required once the plant becomes established, with 1 being the easiest and 3 being reserved for plants that have a few special needs.

We have grown all the fruiting plants in this book at the arboretum, and we have included a "Taste Profile" for each chapter—because for us it all comes down to how a fruit or nut tastes when you decide what to plant in your garden. We hope we'll inspire readers to try a few of these edible fruits that are almost never encountered in supermarkets.

General Considerations

Every site is unique, whether it's a backyard, front yard, raised bed, community space, rooftop garden, farm plot, or container garden on a deck, and there is no substitute for knowing your site. Sometimes you can do lots of research on the growing requirements of a particular fruiting plant and still have it die after lavishing lots of attention on it. If there is one piece of advice we can give, it is this: Gardening takes patience and requires close observance of your landscape over time. But sometimes it is not feasible to be patient, especially when you want to immediately get plants in the ground. We often say that you must kill a plant several times to learn how to grow it correctly. The goal of this book is to expose you to low-maintenance, pest-resistant plants that should require little care after you provide enough water for them for one growing season. We have listed a few things to keep in mind before planting and some suggestions for caring for your plants during the growing season. There are many superb books specifically written about planting, and we have included some of our favorites in appendix 1.

Choosing a Planting Site

When selecting the best planting site for your plants, there are a number of items to consider.

Determine the Soil Content at Your Planting Site: Does the soil at your location contain a heavy amount of clay, sand, or stone, or is it loamy? Depending on how large a planting area you have, there could be many different types of soil on your property. Taking soil samples from several of the places where you want to plant can help determine this. Many good sources of information on soil types can be found in university agricultural programs.

The pH (Acidity) Level in Your Soil: Another factor to take into consideration is the pH (acidity or alkalinity) of your soil. Most plants can live in the slightly acidic range from 6.5 to 5.5 pH. But certain plants require nutrients that can only be absorbed into the root system based on specific pH levels in the soil. Soil testing is one of the best ways to determine what type of soil you have. You can do this by contacting your local university agricultural extension service or the agricultural testing labs in your area. These organizations can sometimes provide soil testing services, in addition to selling simple soil testing kits with instructions that allow you to test your own soil, and can advise what to do if you need to change the acidity levels of your planting site.

Soil Depth in the Planting Area: It is also a good idea to understand how much topsoil you have to work with. Some planting areas have deep soil with a depth of 5 feet (1.5 m) or more before you hit bedrock. Other areas might have shallow soil with only 1 foot (31 cm) to work with. Sometimes digging test holes in several places may help you determine the depth of your soil in order to understand the best place to plant trees on your property. If you have lots of choices about where to plant, choose a site with deep soil, though most plants are adaptable and will tend to have horizontal root structures if they are planted in shallow soil areas. If you have soil that is only a few feet deep, it may benefit your trees to plant them in large spreading mounds. These can be 6- to 10-foot (1.8–3.1 m) wide circles that gently rise to a 1-foot-tall (31 cm) slope. A plant is buried in the center of the mound so that its roots will have more room to spread out. The only drawback to this planting method is that every few years, you will need to add soil to the width of the mound to give the roots more room as they grow outward.

Determine Your Light Exposure: The amount of light that shines directly on a planting area determines the light conditions, which can range from full sun to full shade. This is referred to as a planting aspect, and it will help you determine what type of plant will thrive in a particular site. For the best results locate a planting area that has a large amount of south-facing sun exposure, which is easy to locate with a compass. Following is a general guideline for evaluating the different aspects of sunlight exposure to determine what plants will grow well in a particular planting area. If you only get three hours of sunlight but those are in the hottest part of the day (noon until 3 P.M.), that may be better sun exposure than four hours of sun in the cooler part of the morning. Also keep in mind that the intensity and amount of light exposure changes at different times of the year.

Here are some general guidelines for sun exposure:

Full Sun: 12 to 8 hours of direct sunlight.
Part Shade: 6 to 4 hours of direct sunlight when the sun is hottest (midday to afternoon).
Part Sun: 4 to 2 hours of direct sunlight, generally in the early morning or late day.
Full Shade: No direct sunlight; filtered sunlight to complete shade.

Determine Your Hardiness Zone: Most plants are rated by their hardiness factor, which is defined as the ability of a plant to survive the coldest temperatures of the winter. In addition to the cold, it measures a plant's ability to tolerate heat, drought, flooding, and wind. A hardiness zone is a geographic area defined to encompass a certain range of climatic conditions relevant to plant growth and survival. The USDA Plant Hardiness Zone Map is the standard by which gardeners and growers can determine which plants are most likely to thrive at a location. The map is based on the average annual minimum winter temperature, divided into 10°F (6°C) zones, which was last updated in 2012.[1]

Microclimate: The microclimates in your planting area can also play an important role in a plant's ability to survive. A microclimate is a local set of atmospheric conditions that differ from those in the surrounding areas. This can be as little as a few degrees, but may be substantial enough to allow a less hardy plant to grow in an area where it normally would not survive. If you live in an urban area, hard surfaces like asphalt and concrete absorb the sun's energy and can make a planting environment hotter. South-facing areas are exposed to more direct sunlight and can be warmer for longer periods of time. Also keep in mind that cold air tends to sit in the lowest parts of a planting site, so those areas can stay colder for longer periods of time. Those spots may be more difficult to grow a plant than in a warmer site only 30 feet (9.1 m) uphill from that location.

Determine Your Water Source: In the first year a tree or bush is planted, it will generally need 1 inch (2.5 cm) of water a week to stay healthy. So another important factor that goes into considering where to plant must also be convenient access to water. For watering considerations see the "Planting and Care—The First Season" section on page 11.

Buying a Plant

What you grow can depend on where you are able to purchase it. Your local nursery may have a good selection of different large-sized containers of blueberry varieties, for example, to choose from. But if you are looking for a "Seedless Female" che, it is most likely that it will be available only as a small-sized plant from a specialty mail-order nursery. Because there are several options for buying plants, here are the most common.

Bare-Root Plants: A wide selection of bare-root plants is available through mail-order nurseries in early spring or late fall while the weather is still cool. These plants are dormant and shipped without any soil around their roots. Dormant plants are commonly wrapped with damp shredded paper and enclosed in a plastic bag to keep them moist during the time they are in passage. The advantage of this is being able to see all of the plant's root system. When you place a bare-root shrub or tree in soil, you can arrange the plant's roots so that they will grow into the soil in a symmetrical manner. The disadvantages of buying bare-root plants is that they are generally smaller than what you can buy at a local nursery. Because they have exposed roots, they are more fragile and need to be planted fairly quickly after you receive them. They can easily dry out and must be protected from direct sunlight until planted. Also, the window for shipping bare-root plants is much shorter than for potted plants because it must be done when the weather is cool so they will remain dormant.

Potted Plants: Plastic containers are the standard way nurseries sell plants to the general public. The advantages of potted plants are that they are available for sale for most of the year (excluding winter) and their root systems tend to be more substantial and less likely to be shocked when planting. The downside to a containerized plant is that you cannot see the entire root system, so you may get a plant that is damaged or one that is pot-bound. Also, nurseries tend to stock the familiar, tried-and-true potted plants, so the choices can be limited.

Balled and Burlapped Plants (B & B): These are plants dug up from the ground by a tractor and wrapped. *Balled* is in reference to the shape of the root ball with the soil surrounding it. *Burlapped* refers to the wrapping material that surrounds the root ball and holds the loose soil and roots together. These types of plants are generally more expensive and have larger root systems that will not fit into plastic containers.

FIELD GARDEN

Some people are willing to pay the high prices for these extra-large plants because they want to get an instant big tree on their planting site. We believe that there are not enough advantages to buying B & B plants and generally advise against them. Larger specimens often weigh hundreds of pounds, which is another disadvantage, and have a wire metal basket around the root system to contain all the loose soil, and many landscapers do not remove the burlap or metal cages from the plants before putting them in the ground. This bad, old-fashioned gardening practice can seriously damage a plant's growing root system; we have seen old plant roots girdled and strangled by these metal cages.

Plants maintain a growth ratio between the roots and the growth aboveground, so the more intact a plant's root system is, the quicker it will put vigorous energy into the growth above the ground. Although it is tempting to start with a big plant, a study by the Cornell University Agricultural Extension shows that common B & B plants have over 90 percent of their roots cut off when they are mechanically dug up. In another study by Cornell, two sets of the same species of trees were planted in one location. One set were moderate-sized nursery trees with intact root systems and the second set of trees were large B & B trees that had lost 90 percent of their roots. Over each growing season the smaller trees with intact root systems grew larger, while the B & B trees with reduced root balls only grew a few inches each year. After thirteen years both sets of trees were about the same size![2]

WHAT TO LOOK FOR WHEN BUYING A PLANT

Selecting a fruiting plant that has been properly grown and shaped by a nursery is the first step toward planting a tree or a bush that will thrive and be productive for many years.

The Structure of the Tree: Commercial growers that supply plants for nurseries often cut off the lower limbs of fruiting trees to raise the crown and create a taller tree structure, because most nursery customers that purchase a taller tree think they are getting a better deal. Under these circumstances the grower has shaped a tree for tallness but not for easy fruit picking. Picking fruit at a height of 4 to 7 feet (1.2–2.1 m) is more convenient than climbing on a ladder to get fruit that is 6 to 12 feet (1.8–3.7 m) high. So purchasing a smaller tree that can be pruned into the tree you ultimately want may be better than getting one already pre-shaped for height.

There is one positive advantage to having a tree with a higher canopy: Higher branches are less likely to be chewed upon by deer. This may be a good strategy if you have heavy deer activity in your planting area with no protective fencing.

Disease-Resistant Varieties: A plant's disease resistance is its genetic ability to prevent a disease-causing pathogen from harming it. Some species develop resistance to diseases through deliberate plant breeding or through the process of natural selection. In addition to good fruit productivity, many modern varieties have also been bred to resist the effects of different types of pathogens. When choosing plants, look for varieties and cultivars that have been bred to have the most resistance for diseases known to be prevalent in your geographic area.

Planting and Care–The First Season

The general rule of thumb for planting a tree is to dig the hole two to three times wider than the pot that the plant has been grown in, and to place the plant in the hole so that the top of the root ball is level with the surface of the soil. Nursery plants often have root systems that are tangled together from spending many years growing inside a plastic pot. Depending on how root-bound the plant is, it may be beneficial to loosen up some of the finer roots, and then gently pull them away from the outermost part of the root ball, to encourage them to grow outward into the soil.

There are lots of different theories about the soil that should be used to fill in the planting hole around the plant. Because the new plant will ultimately grow into the pre-existing soil at your planting site, we believe in using the original soil from the hole, amended with a mixture of compost, well-decomposed leaf mulch, or at the very least organic garden soil. Since each location is unique, there is no single answer about how to amend soil except to avoid putting heavy fertilizer directly against the root ball. Heavy amounts of any product with lots of nitrogen can actually burn up a plant's roots. Do your research on the various requirements for each plant in order to create the proper conditions for it to thrive. There are many fine books and articles that go into more depth on these practices, such as the ones listed in appendix 1.

When to Plant a Tree: With the exception of winter, when the ground is frozen, you can plant a tree any time of year as long as you are going to consistently water it. Spring and fall are the parts of the growing season with the coolest daytime temperatures, making them the optimal times to plant. Cooler weather allows moisture to remain in the soil after watering so that you may not need to worry about the roots of the plant drying out. If a drought occurs the second year after planting a tree, it may be necessary to keep watering the plant in order to keep it healthy. This is especially important for small plants sited in full sunlight.

Watering Plants the First Year: It is important to make sure that your plant receives a consistent amount of water for the first growing season—this is the single most important factor for its survival. A good amount of water the first year, and sometimes the second, will keep a new tree or bush healthy and give it the best opportunity to become a long-lived, productive plant. If you put your fingers in the soil and the first inch or two (2.5–5.1 cm) is moist, your plants can probably forgo watering for a day. During the middle of summer when daytime temperatures rise, new plantings may need to be watered every few days (or every day) if

there hasn't been any rain. Generally speaking, a new plant needs to receive 1 inch (2.5 cm) of water per week, and during the summer, *rain will not be a reliable source of moisture*. Getting water to your plants is important and easily overlooked, so once you determine the source for water, you have to decide how the individual plants will be cared for. If you are hand-watering more than a few trees, you have to be realistic about getting water to them if they are spread out across a large area, so make sure you have a reliable source of water for your new plants.

Mulch Holding Moisture: A 1- to 2-inch (2.5–5.1 cm) layer of organic mulch will keep the roots of young plants moist longer than bare soil. During the growing season, wood mulch will not only help a young plant's roots to stay cool and moist, but also help suppress weeds from growing near the plant. Try to keep the mulch a few inches away from directly touching the plant's trunk and spread it no more than 2½ inches (6.4 cm) thick, because bacterial diseases can live in thick piles of mulch.

A Variation on Using a Hose: Most people just turn on a hose to water their plants. This works fine, but when you stand above a plant and spray it with a hose, much of the water is wasted by being dispersed sideways away from the plant and not directly toward the roots. A good strategy for newer plantings is to turn a faucet 90 percent off and allow just a trickle of water to run through the hose. If the end of a hose is placed right near the plant's trunk, the water will slowly seep downward directly into the roots, encouraging them to grow deeper into the soil. A hose can be left on for half an hour like this and barely waste any water. If this is done several times a season, it will build a stronger root system on new plants.

Gator Bags and Watering Bags: These products allow plants to be watered in the same manner as the "slowly trickling hose" method mentioned above. There are two basic types of commercially manufactured watering bags used for new plantings. Gator bags are watering bags that are zipped like a vest around the bottom of a tree trunk. These bags have an open pouch that is filled up with water. The chief benefit of using one of these bags is that the bottom of the bag has tiny pinprick holes that permit water to slowly trickle out. A Gator bag is 16 inches (40.6 cm) in height and holds 20 gallons (75.8 L) of water, which will water a tree for hours, deeply saturating the roots. They are best reserved for trees and plants with long trunks. Watering bags function the same way as

Gator bags, but are flat and doughnut-shaped. These circle the bottom of a trunk, which is convenient for slipping them around shrubs and trees with low branches.

Drip Irrigation Systems and Timers: These are watering systems that deliver water from a water faucet to your plants so that you do not have to. They range in scope from a simple battery-operated timer hooked up to an outside faucet, to the complicated types of watering systems that connect to an electrical panel and read the amount of moisture in your soil and water the plants accordingly. A basic battery-operated timer hooks up to a faucet and connects to plastic tubing that runs to the plants to be watered. The timer opens up the faucet at pre-programmed periods of time, which allows water to run through the tubing to the plants. Small sprayers or emitters are attached to the end of the plastic tubing and permit the water to flow to the root systems of plants. This type of system and the more complicated watering system mentioned above are available in a range of styles through specialty mail-order nursery catalogs.

Amendments

Listed below are a few of the basic organic fertilizers and soil amendments for enriching the soil that surrounds new plantings. Most fertilizers should be applied over the topsoil of new plants in moderation on an annual basis. Plants tend to grow best when they receive nutrients that can be absorbed slowly over a period of time and not applied as a single large load.

Commercial Organic Fertilizers: Many commercially manufactured organic fertilizers are available at garden centers and hardware stores. These fertilizers are made from natural products and elements such as seaweed, magnesium, bonemeal, copper, and potash. Many of these products are beneficial to new plantings but should be used in strict accordance with the manufacturer's directions and applied safely with protective gloves.

Compost: An old-fashioned organic way to get a natural fertilizer is to use decomposed vegetable food waste and other organic decomposed material after it has broken down for a sufficient period of time. Compost will provide lots of natural minerals, add beneficial bacteria, and enhance the texture of the soil that your plants are growing in.

Compost Tea: Compost tea is made by placing compost in a plastic bucket and filling the bucket up with water. After a period of time, the water can be separated from the compost by pouring it into a new container. The water will then contain the nutrients of the compost and can be used as a liquid fertilizer for most garden plants. You can also make a tea using well-aged manure, wood ash, and seaweed.

Gravel: Small amounts of gravel can be used for stabilizing raised planting mounds and to provide drainage.

Leaf Mulch / Leaf Mold / Leaf Humus: These terms are interchangeable for our favorite soil fertilizer. It's a nutrient-rich amendment produced through the fungal decomposition of tree leaves, and it's free! Deciduous trees store most of their nutrients in their leaf canopy, and those nutrients get dispersed to their root systems when the leaves drop to the ground in the fall. This annual process slowly adds valuable organic matter to the soil around the trees. This natural conditioner improves the water-holding capacity of soil, enhances soil structure (and therefore water and air movement within the soil), and provides a habitat for micro- and macroorganisms.

If you are growing fruiting plants, it will be well worth your time to begin collecting and storing fall leaf-litter in a large pile. After a few years you can dig down through the side of the pile in order to take away the broken-down leaves, which will be a rich dark brown color that looks more like soil and less like the original leaf litter. This leaf mulch can then be used as a soil amendment to create fertility as well as improve the texture of the soil.

Manure (Chicken, Cow, Horse, Alpaca, Etc.): Well-aged manures that are one year old or older are rich in plant nutrients such as nitrogen (N), phosphorus (P), and potassium (K), and provide organic matter that conditions the soil and improves its water-holding capacity. Manure typically comes mixed with animal bedding such as sawdust, wood shavings, or feed waste that also adds nutrients and fertility to the soil. Each type of animal manure has different chemical properties, so it's best to contact your local agricultural extension service for guidelines on using them.

Pine Leaf Mulch: The cast-off needle-like leaves from conifers are a great landscaping mulch and acidify the soil as they break down, which provides an excellent amendment for acid-loving plants such as blueberries and Korean stone pine.

Wood Ash: When wood is burned it gives off nitrogen and sulfur as gases, but leaves behind potassium and other plant nutrients in small amounts. This is a good soil additive and raises the soil pH, making it more alkaline, which is the opposite of what pine mulch does. Some fruiting plants such as gooseberry and currants benefit from a handful of potash each spring.

Delayed Planting and Storage Containers

Ordering from mail-order catalogs will allow you access to hundreds of different types of food plants that you will not see for sale at local garden centers. Mail-ordered plants are generally smaller than what you can purchase in a local nursery, and while we believe in supporting local, family-owned nurseries, we also feel that having access to a wider selection of rare fruiting plants is important for biodiversity.

Delayed Planting: Many people purchase a small plant or seedling feeling confident that after planting it, they will be able to water it on a regular basis for the first season. We have seen many small plants and seedlings killed this way. We think it is prudent to delay planting, especially if it is really small, expensive, or purchased through a mail-order nursery. Keeping these smaller plants in pots for a few years in a storage container before actually planting them may help prevent their root systems from drying out during their first growing season. A plant in the ground with a 3-inch (7.6 cm) root ball that is exposed to a long, hot summer does not have as much resiliency and chance of surviving as a plant with a more mature root ball. Temporarily storing and watering plants inside a protective storage container until the roots grow larger may increase their chance of survival.

Storage Containers: A storage container is useful to protect potted plants and minimizes their water loss during the growing season. This can range from a set of potted plants placed together in one larger plastic pot covered over in mulch, to a constructed wooden box with an open bottom, filled with pots that have mulch over them to protect them against direct exposure to the sun. The purpose of a plant storage container is to make the plants feel more like they are planted in the ground. The small potted plants can be placed inside the container together, and mulch is used to fill around the pots up to the edges of the container. This allows all

the plants to be watered at one time and kept moist in lots of sunlight. You can even forgo a formal storage container by digging a hole in the soil and burying the potted plants directly in the hole. Placing mulch over the pots will help their roots stay cool.

Pest Control and Animal Deterrent

We have chosen these fifty fruit and nut trees because of the relative ease of growing and maintaining them as productive plants. However, all plants are susceptible to a certain amount of insect and animal pressure, whether it is the small larva of an elderberry borer beetle that chew holes in stems in the early summer, or field mice that strip and eat the tender bark of young trees under the snow in winter. Both young and mature trees can be susceptible to a variety of different pest problems. Below are some strategies that we have found to be helpful.

Dormant Oil / Horticultural Oil: Dormant oil, also known as horticultural oil, is made from refined petroleum oil or vegetable oil, and its viscosity makes it effective in killing pests by smothering them. It is best applied in the late winter when trees are dormant (inactive), and in early spring when temperatures are just above freezing. The oil kills overwintering insects like aphids, mealybugs, thrips, whiteflies, adelgids, caterpillar eggs, leafhoppers, scale, and mites. Horticultural oil is more effective on young insects and the eggs deposited on plant bark but less useful on mature insects, so good timing and early intervention are important. The oil dissipates and doesn't leave behind a toxic residue, which makes it ideal for use on fruiting plants because it will not affect the pollinators that arrive later in the growing season. It is also considered safe to use around humans and pets.

Surround WP: Surround WP is an organic pesticide made out of finely ground kaolin clay, so it is a fairly mild, organically approved pest control. Surround WP discourages insects from laying eggs on fruit by creating a physical barrier that irritates and confuses insects trying to crawl or feed on the leaves and the fruit, but does not inhibit the fruit's ability to grow. Once the fruit is picked, Surround WP can easily be washed off. The drawback to using Surround WP is that it turns the leaves and fruit a filmy white color, which does not harm the plant in any way, but some people may find it aesthetically unpleasant. Also rain can dilute the clay layer protecting the fruit, so you will have to use additional applications of the spray after heavy rainstorms.

PROTECTING YOUNG TREE TRUNKS

In suburban and rural settings, many people think that deer are the greatest threat to their plantings, but actually voles, field mice, and rabbits are more likely to kill your plants than a deer. These animals are voracious eaters of vegetation, and during the warm part of the year, they have an infinite amount of wild vegetation to choose from and so are less likely to harm new fruit trees and shrubs. The problems tend to occur in winter when snow coats the ground and these animals run out of food options. Voles and field mice make burrows in the ground and tunnel under the snow to move around the landscape hunting for food. It is then that a young shrub or tree's thin bark becomes an appealing food choice. The parts of the plants located near the ground that are covered by snow can be reached by small rodents that use their teeth to scrape the bark off plants, and rabbits can chew on tree trunks above the snow line. If a continuous strip of bark is chewed in a circle around a plant's trunk, it will die, because the water and nutrients from the soil will be cut off. As trees mature their outer bark grows thicker and "corkier," and animal teeth have more difficulty cutting through coarse plant bark.

Providing a protective wrap for the lower sections of a plant's trunk for several years will help the plant survive. We stop worrying about tree trunk protection when the plants have grown thick bark near the bottom of the tree, because animals cannot chew through it. In the case of shrubs, a small animal probably won't be able to eat the bark of every branch if there are a large quantity of thick ones.

Fortunately, there are several different types of plant protection. Each of these methods has advantages and disadvantages.

Plastic Vinyl Spiral Tree Guards: These are stretchable white plastic strips that wind around the trunk of a tree to cover it from the ground up a couple of feet (60 cm) or so. This layer of rigid plastic prevents small mammals from chewing on the trunk under the snow and stops rabbits from chewing on a tree trunk above the snow; it can also protect the tree from winter sunscald. These white vinyl guards are often already wrapped around tree trunks when you buy them from a nursery, but you can also purchase this product through mail-order catalogs. The only drawback is that people tend to leave them on for the entire season, if not for years. During the warm part of the growing season, insects can use the inside of the tree guard as a place to hide and deposit eggs before burrowing into the trunk; then these vinyl strips become a liability because they provide a hidden nesting site. Our advice is to remove these tree guards early in spring, and then put them around the trunks again in late fall as the trees go dormant.

Plastic Mesh and Plastic Tubing Trunk Protectors: Another way to protect your trees from animal damage in winter are molded plastic trunk protectors that usually come in lengths of 2 to 6 feet (0.6–1.8 m) and are available through mail-order companies. These tree trunk wrappers have an opening running down the length of one side of the plastic to allow the guard to be circled around the trunk of a plant and secured with wire. There are generally two types. One comes in the form of a plastic fence with square openings in a grid, and the other an expandable solid piece of circular plastic tubing. Each of these trunk guards can be wrapped around young vulnerable trees for a few years, then removed after the plants have reached sufficient maturity and have grown thicker bark. The biggest drawback to using the solid

plastic tubing around seedling trees is that you cannot clearly see the bottom of the tubing, which means that insects and voles may use them for nests, so remember to periodically open them up for inspections.

Hardware Cloth Wrappings: This is our preferred method of guarding against the winter damage of voles, field mice, and rabbits. Hardware cloth comes in rolls that are widely available in hardware stores and home improvement centers. These metal rolls are in the form of a screen made from thin wire manufactured into a grid of ¼-inch (6 mm) squares. It can be easily cut with tin snips into various lengths. A cylinder of cut hardware cloth is circled around the lower trunk of the plant and held in place with wire. Since you can see through hardware cloth cylinders, you can leave them on the tree for several years as long as you inspect them periodically and have left sufficient space to accommodate for the plant's trunk growth. After several growing seasons, as the trunk gets larger and develops thicker bark, you can take off the guard and reuse it.

Green Metal Fencing for Trees: If you are only planting a few trees and there is heavy deer activity around your planting site, building a circular protective tree cage out of metal fencing may be a good solution. Most home improvement centers and some hardware stores sell rolls of metal deer fencing that come in heights of 3 to 6 feet (0.9–1.8 m), in rolls that are 50 feet (15.2 m) long and painted with a durable black or green plastic paint. Cut sections of fencing are formed into a large circle with the two ends tied together so that the whole piece of fencing feels like a sturdy enclosed circle. This large circle can be placed around a young tree, with stakes hammered into the ground to keep it in place. These metal circles are a good protective solution for open planting sites, and after four to seven growing seasons the trees will develop thick trunks and grow taller than the reach of most deer. At this point the deer cannot completely destroy a tree when they chew on its lower branches, so you can remove the metal circles and use them on other trees.

Professional Polypropylene Deer Fencing: The next step up from circular metal cages is to completely surround your planting site with an enclosure of durable plastic deer fencing. This is expensive and time-consuming but can last for two decades, if not longer, with periodic maintenance. Professional deer fencing is worth the effort, and if you are going to plant a large number of trees and shrubs over a sizable area, we believe that you should use heavy-duty professional-grade fencing. Higher-quality grades of deer fencing have to be mail-ordered from specialty fencing companies, which offer several different styles. We recommend a sturdy type of fencing made out of polypropylene plastic, which is manufactured in 6- to 8-foot (1.8–2.4 m) heights in rolls of 100 to 300 feet (30.5–91.4 m). The fencing is made out of thick, durable black plastic formed into a grid of 2-inch (5.1 cm) squares. The most durable grades of this fencing are rated to take 650 pounds (295 kg) of impact per square inch. This type of protective fencing is held in place by attaching it to metal poles, cedar-log posts, treated 2×4s, or trees in a forest setting. When this type of fencing is attached to trees in a forest, the black plastic blends into the shadows of the trees and cannot be seen from a distance, which is one of its best features.

Metal Deer Fencing: There are several grades of commercial metal deer fencing manufactured in 4- to 8-foot (1.2–2.4 m) heights and sold in large rolls. Such fencing is available through specialty mail-order. Metal rolls are more expensive than the polypropylene rolls but may be worth the expense if you want to put up one fence system that will last your entire lifetime, which rabbits and woodchucks can't chew through.

Winter Protection

Anyone who has tended a garden in the northern parts of the United States over several years understands that each growing season is distinct because of the vagaries of the weather. While each season has similar qualities to those of the year before, no two years are exactly alike. Some winter seasons bring massive amounts of snow and abnormally cold weather, while others deliver only a few inches of snow. All the combined elements of snow, wind, temperature, sun exposure, and fluctuations between winter and spring affect the plants in a garden. Those readers who live where winter temperatures do not fall below 10°F (−12°C) might not need to worry as much about the winter protection measures listed below. When a tree or shrub is young and has only been in the ground for a single growing season, it is vulnerable to winter damage. While some plants are tough and adaptable, others are more fragile and may need a few years of extra protection to survive the winter. The suggestions below are a range of options that you can employ for a single tree or for multiple plantings that would benefit from a year or two of extra protection before being exposed to the full impact of a northern winter.

Many may find that lavishing extra winter care on new plants is excessive. The degree to which you protect a young planting that may not be completely hardy can range from doing nothing to constructing a small temporary enclosure. We have listed some options below ranked from the simplest to the most complex.

Protective Mulching: The easiest way to reduce the negative effects of low winter temperatures on a young plant is a thick layer of mulch over the root system. Any mulch of mixed wood chips or shavings will cushion a young plant's roots against winter temperatures. This extra mulch can be piled high against a plant's

Hardiness and Marginal Plants

Rating a given tree or shrub's chances of surviving in any given environment depends on a whole host of factors. Is it a small three-year-old seedling, or a large, established bush with a dense root system and thick bark? Is it only hardy when its roots are dry, or can it take sitting under a foot (31 cm) of snow for long periods of time? Is the plant cold-tolerant but intolerant of an open exposed planting site that is windy? These considerations must also be factored against the backdrop of how each winter's season affects a particular plant. A marginally hardy plant is one that is close to being hardy but not able to survive the worst of your winter's conditions. An extremely hard winter may kill a young tree while a milder one produces only minor damage. In such a case, you could protect a tree for a few years to allow its root system to get larger and its bark to get thicker. When the difference between a plant's survival is a matter of a few degrees, it is possible to push the hardiness zones. Also important for a marginal plant's survival is the time of year that you plant it. A tree planted in early April rather than September will have more time for its roots to grow into the soil and become more robust before winter. If you plant a new shrub that is much more cold-tolerant than your local growing conditions, these issues will not be as much of a concern.

trunk or branches for protection as an insulation blanket over the root system. Then in spring the excess mulch must be cleared away from directly touching the trunk and spread out evenly in a thin layer in order to cover the roots. Because thick layers of mulch provide cover for small rodents, it's important to use a protective tree collar so that animals cannot damage the new plant.

Winter Sunscald and Painting Tree Trunks: Woody deciduous plants survive the winter months by going into a period of dormancy—the state in which a plant is alive but not actively growing. This is triggered by a combination of the autumn cold and the shortening day length. Toward the end of winter, as daytime temperatures begin warming up, deciduous plants can sometimes start to wake from their dormancy period while the land is still in the grip of winter, and this can cause damage. Warm daytime temperatures cause the frozen sap near the roots of a woody plant to rise up the trunk toward the buds on the ends the branches. Because it's winter, though, nighttime temperatures can still drop below freezing. The sap that has risen up the trunk can

suddenly freeze before it has enough time to travel back down to the roots. When the temperature warms up again, the trapped frozen sap expands, splitting the trunk open under the pressure. This is called *southwestern syndrome* or *sunscald*, and it is exemplified by cracks that run down the bark of a plant on the side with the sunniest exposure, typically facing the southwest.

Sunscald is most prevalent in young trees with long trunks that contain thin, flexible bark that are planted in full-sun sites facing a southern or southwestern exposure. A traditional method for dealing with this is by painting the trunks of young trees with a thin layer of white paint diluted with water. The theory is that the young plant's white-painted bark reflects sunlight away from the trunk, slowing the warming of the tree's sap. Take water-based white house paint and thin it with water at a ratio of 1 part paint to 4 parts water, then apply a few thin coats of paint from the tree trunk's bottom to the first major branches. The primary drawback is that the tree's bark will look chalky white and less aesthetically pleasing. However, as the plant matures the paint slowly fades, and splitting tends to happen less as the tree develops thicker bark.

Wrapping Plants: An old-fashioned way of protecting plants is to wrap them with burlap. This will provide some protection from the cold, but is best used at planting sites that have extremely windy conditions where the branches of plants can be damaged by desiccation (drying out). This method of protection can also be used to protect shrubs planted closely together in a hedge. Wrap the burlap firmly around the sides of the hedge as a large protective layer against heavy snow loads.

Tenting: You can construct a temporary tent around a shrub or small tree to create a mini greenhouse environment through the winter. A protective tent enclosure can be a simple "tepee," formed out of thin wooden posts or bamboo stakes driven into the ground and tied together on top. You then wrap a transparent plastic sheet around the entire structure. Remove the covering in early spring before the temperatures get too warm. Remember that any warm protective space for a plant is also a warm protective space for small mice and voles, which may chew on the lower sections of the plant. If mice and voles are a problem in your region, consider putting a protective vole cover around the plants.

Greenhouse: The largest form of winter protection for growing plants in the cold northern latitudes is a greenhouse / hoop house.

Because a greenhouse interior can be 10 to 30°F (5–17°C) hotter than outside temperatures in the daytime, a greenhouse lengthens the growing season. In addition to winter protection, a greenhouse cushions a plant against the wind, snowfall, and icy conditions. Although greenhouses are expensive to build, they can last for decades and allow you to consistently produce food over a longer growing season. Commercially produced greenhouse kits are available in a wide range of sizes and styles. Because the environment in an unheated greenhouse is warmer than the surrounding landscape, plants that are marginal (one zone warmer than your region) can often be kept alive and produce fruit. Although there are size limitations to growing plants in a greenhouse, fruit trees can survive and produce fruit there through diligent pruning and management.

Akebia

Akebia quinata, A. trifoliata

Chocolate Vine, Five-Leaf Akebia, Three-Leaf Akebia

> *. . . The fruits . . . are even more fantastic than the flowers and look like chalky bananas of pale blue leather. The blue has to be seen to be believed: it is not a shade common to fruits . . . it is a brilliant violet or indigo, softened by a delicate gray bloom, and is altogether more reminiscent of a suede dancing pump or a fancy coin purse than it is of flowers or fruit.*
>
> —EDGAR ANDERSON, 1934[1]

As readers might gather from the description of akebia fruit in the epigraph, akebia is just plain strange. This wonderful plant produces fruit that—more than any other fruit in our collection—looks like it dropped from outer space! When visitors to the gardens see the fruit, the overall effect is "*Wow!* What the hell is that?" Because of their attractive foliage and delicate flowers, these plants are occasionally sold in the nursery trade as hardy, fast-growing, and shade-adaptable vines. They are useful for readily scaling walls, fences, and trellises, or spreading out to cover empty garden areas as a ground cover. Because they are only marketed as "ornamental vines," the edible aspects of akebia are usually ignored by nurseries. So most gardeners do not know that you need two genetically different akebia plants to get the strange, distinctive fruit. As a result, gardeners often have only one akebia vine and may have no idea this vine produces fruit.

Akebia belong to Lardizabalaceae, a small family of plants that includes vines and trees originating in China, Japan, and Korea. Akebia has been cultivated for hundreds of years in Asia for the edible fruits and for medicinal uses.[2]

In Japan the akebia is sold for a high price in large supermarkets, and the fruits are referred to as *Princess of the Mountain*.[3] In Japan the fruit is called *akebi*, which is a compound word combining "open" and "fruit," a reference to the fruit's seedpods splitting open when ready to eat. Japanese culture even distinguishes two different types of ripeness. The term *Murasaki Akebi* is used when the seedpods have reached edibility but are still a bright purple color, and *Ishi Akebi* (which translates as "stone akebi") when the outer skin on the fruit turns a tobacco-brown color before splitting open.[4] *Akebia quinata* was introduced to England in 1843 by the great plant explorer Robert Fortune, who discovered the vines on the island of Chusan, where he saw the plant vining up trees and hanging down off their branches in "graceful festoons."[5]

Akebia is noted for its fast growth, and we have heard other gardeners say that this vigorous vine can be extremely invasive.[6] But in our eighteen years of growing many different cultivars of akebia, we have seen them grow aggressively, but not invasively (see the "Site and Soil Conditions" section below).

Growth Difficulty Rating: 1. Akebia is easy to grow and thrives on neglect once established.

Taste Profile and Uses: The fruit is 3 to 5 inches (7.6–12.7 cm) long and sausage-shaped, with a rich violet-purple color covered in a waxy bloom. When fully ripe, the fruit splits open like a coin purse to reveal tiny black seeds suspended in creamy, translucent white pulp. The pulp is mildly sweet, though some might find it bland with a taste that is most reminiscent of tapioca pudding. The seeds can be a little sour, so we tend to eat seeds and pulp together without chewing the seeds. Though not overwhelmingly strong in flavor, the overall effect is like eating tapioca lightly flavored with coconut.

Plant Description: This large woody, deciduous, rambling, twining vine has small, lightly scented flowers that bloom in spring after the leaf buds have opened up. Both species of akebia can produce foliage that grows into thick masses that sometimes obscure the small, delicate-looking flowers. Five-leaf akebia (*A. quinata*) produces foliage in clusters of five oblong-shaped leaves that are 1½ to 3 inches (3.8–7.6 cm) long, and mature vines grow up to 35 feet (10.7 m).

Three-leaf akebia (*A. trifoliata*) has foliage arranged in groups of three with a wavy leaf margin with leaves that are obovate and more circular-shaped than *A. quinata,* growing 1½ to 4 inches (3.8–10.2 cm) long. It grows less vigorously.

Flowers: From a distance the akebia flowers might go unnoticed because the dense foliage of the vine often covers the delicate blooms, but up close both species have beautiful flowers that are interesting because they are apetulous (have no flower petals), so the “flower” is actually composed of the sepals (modified leaves).

Young fruits developing.

The beautiful flowers.

Ripe akebia fruit cluster.

The foliage of the five-leaf and three-leaf akebia.

Akebia vines on a cedar structure.

Five-leaf akebia has female flowers that appear on racemes (flower clusters) in groups of two. They are 1 to 1½ inches (2.5–3.8 cm) in diameter and are composed of three chocolate-purple, fleshy, concave sepals that surround a group of cylindrical pistils. The smaller male flowers are light pink in color and produced on the upper part of the flower stem with recurved sepals, and are roughly ½ inch (1.3 cm) or smaller in size.

Three-leaf akebia has a similar habit, and the flowers are slightly smaller and an attractive reddish-purple color. There are also several hybrid flower colors available in the nursery trade (see the "Cultivars" section below).

Pollination Requirements: The literature on the pollination of akebia flowers is confusing and contains contradictions involving too much detail to cover here. In general, akebia vines contain a raceme, or cluster of female flowers, with a separate male flower arising from the base of those females. But vines can produce flowers with both sexual attributes. To add to this confusion is the fact that female flowers generally open up before male flowers on the same plant and that vines are self-sterile and will not pollinate themselves. This all gets botanically muddy, but horticultural literature and nursery catalogs agree that in order to produce fruit, you need at least two individual vines that are not the same plant: that is, two different cultivars of the same species or two different species of akebia. Because akebia flowers open up over an extended period of time, two or more vines will allow more opportunities for cross-pollination between different flowers on separate plants.[7] Hand pollination will increase the probability of the vines producing fruit, but these vines are not dependable for setting fruit each season. After many years of growing akebia, we generally get a small amount of fruit each year.

Site and Soil Conditions: This vine is adaptable and tough with a tolerance for various soil conditions and can be planted in full sun or part shade. We feel that this vine is best used in a *part-shade* planting site. While some sources list this vine as being incredibly invasive,[8] our vines have grown in an aggressive manner *but not spread rampantly*. Our akebias are planted in a part-shade aspect (two to four hours of light a day), and this may be one of the reasons why their growth has been kept in check. We have seen a few examples of akebia in full sun where it did grow more uncontrollably, so we advise against planting akebia in full-sun sites.

Hardiness: *Akebia quinata* is rated to USDA zone 4, or −25°F (−32°C), while *A. trifoliata* is rated as slightly less hardy.

Fertilization and Growth: Both are adaptable to many types of soil as long as it is not too wet.

Like all plants native to the forest, akebia would benefit from a yearly addition of leaf mulch or compost.

Cultivars: There are a handful of varieties of akebia that have been selected for different flower colors.

'Deep Purple' (*A. trifoliata*): A deep maroon purple-brown.

'Purple Rose' (*A. trifoliata*): An attractive matte rose-red color.

'Silver Bells' (*A. quinata*): White outer petals and lilac-purple sepals in the center.

Related Species: China blue vine (*Holboellia* spp.) is another genus of attractive flowering vines in the Lardizabalaceae family that produces edible fruit. *H. coriacea* is rated hardy to zone 7. Blue bean tree (*Decaisnea fargesii*) is a medium-sized, multi-limbed shrub or small tree that produces metallic blue seedpods with a clear jelly and lots of seeds inside. They are less pleasant to eat than akebia fruit.

Propagation: This vine is easy to propagate by digging up vine layers, or you can grow seed potted up fresh from the fruit. The seedlings can be grown out for two or three years until they have produced large enough root systems to plant in the ground.

Pests and Problems: No major pests seem to bother akebia. The major problem with akebia is its aggressive growth. We feel that eventually almost any vine will get large enough that you'll have to keep the growth in check by pruning.

Invasive Plant Warnings: The Missouri Botanical Garden website lists akebia as an invasive species in many midwestern states.[9] Check with your local agricultural extension service to check if this vine is a local problem and should be avoided.

Almond

Prunus amygdalus, P. dulcis

Bitter Almond, Sweet Almond

> *No tree of the nut tree group is more exacting than the almond in its demand for just the right condition of soil and climate, but the many varieties will find soils to their individual liking.*
>
> —Robert T. Morris, 1921[1]

The cultivation of almond trees goes back a long way through the history of human civilization, to such an extent that they were widely grown throughout the ancient world. There are references to almonds in the Bible, in China from the tenth century BCE, throughout the Roman Empire and ancient Greece, as well as in ancient Egypt. There are two types of almonds: bitter and sweet. Ancient almonds began in the bitter form—they are a seed from the *Prunus* genus, and like all the seeds of that group they contain a form of cyanide that has to be cooked to remove the bitter and toxic compounds.[2] Over thousands of years farmers have domesticated almond trees and selected the varieties that produce mostly sweet-flavored nut crops. These are now the dominant almonds being cultivated throughout the world.

In the modern world *P. dulcis* is an economically important crop that is primarily grown in Mediterranean climates such as central and western Asia, Iran, Turkey, northern Greece, and North Africa. Almond agriculture spread throughout the world, and it became a major agricultural crop in the United States. In California, almond trees began to be planted sporadically as a crop in the 1850s; by 1909 the Golden State had no fewer than four thousand of these trees planted for commercial production. Currently, California is the major center of almond crops and accounts for 70 percent of the world's supply.[3]

The almond as an edible commodity is widely used as a healthy addition for many meals, added to desserts and made into candy such as marzipan, and processed into a dairy-free milk substitute. Also, after harvesting, the remnants of the shells and hulls are used to feed livestock and to generate electricity in California.[4]

As the nuts mature on the tree, they develop a thin protective covering called a hull that closely resembles a shriveled-up peach. Indeed, almonds and peaches are so closely related that some botanists believe that the two species came from the same evolutionary ancestor and are the result of thousands of years of human cultivation.[5] Botanically speaking, almonds are not "nuts" but are in fact members of the *Prunus* (or stone fruits) genus, which includes cherries and apricots. In other words, you are eating the seeds from fruit trees that have been bred over hundreds of years to reduce the amount of pericarp (fruit pulp) growth in favor of the size of the pit.

The best chance for producing almond crops in the northern parts of the United States is to choose those varieties selected for delayed blossoming, which will not be affected by late frosts. That being said, we have grown two almond trees for over seven years that have survived many hard winters and some late frosts. We think that the best cold-hardy almond cultivars are worth a try if you love the nuts and have space for them in your full-sun planting site.

Growth Difficulty Rating: 2. Although mostly trouble-free, all trees in the *Prunus* genus will require a little extra care in dealing with diseases.

Taste Profile and Uses: Almond is classified as a drupe that consists of a seed surrounded by a hard tan-colored outer shell enclosed in a thin hull that resembles a small green peach. When the fruit is

The ornamental early-spring flowers.

Almonds ripening.

Almond developing after flowering.

A split almond hull.

ripe, the hull splits open to reveal the hard shell that encases the kernel (which is the almond).

Fruit-set occurs on one-year-old wood as well as two- and three-year-old spurs, so older branches should be regularly pruned out to encourage new fruiting spurs. As the almonds ripen on the tree, the hulls that cover their shells begin to split open; this is the indication that they are ready to harvest. This is accomplished by gently tapping branches with a pole or by placing a sheet on the ground and gently shaking the branches to encourage ripe nuts to fall off the tree. After this the hulls must be promptly removed from the shell-encased nuts to prevent mold and to allow the almonds to continue to dry, which improves the texture of the nuts and allows for long-term storage without the danger of mold. Test to see if they have dried properly by cracking open a few of the shells. If the almonds are still rubbery and flexible, give them more time to dry. Once properly dried it is best to freeze the almonds for forty-eight hours to destroy any insects and their eggs. Then they can be thawed for use and placed in a dry container for long-term storage.[6]

Almonds have a distinctive nut flavor and are rich with protein and small amounts of many other vitamins. Because of its mild taste, the almond has an almost infinite number of uses as an addition to both sweet and savory foods. It is rare to find bitter

Unshelled 'Hall's Hardy' almonds.

almond being sold in the nursery trade, but if you plant a bitter almond tree the nuts will need to be cooked before eating to remove the toxic compounds.

Plant Description: The almond is a slow-growing deciduous tree, 15 to 20 feet (4.6–6.1 m) tall, that starts its early years growing fairly upright then begins to branch into a wide-spreading crown as it matures. Almond trees have long, blade-like leaves similar to peach foliage with finely toothed edges. They can be slow to produce nuts and may take up to seven years to yield good crops. Almond trees grown on their own root systems can produce a crop over a long life span, remaining productive for up to fifty years, but most modern cultivars are grafted onto peach rootstocks, which gives them a life span of around twenty years.

Flowers: Early-spring blossoms are highly ornamental and open up before the tree leafs out. The flowers have five large petals that can vary from pink to pinkish white or white in color, with each flower producing attractive, elongated pink sepals. Not all the flower blossoms in a given growing season are capable of bearing fruit, so it is not necessary to thin crops to increase the quantity of the nuts.

Pollination Requirements: Although some varieties are partially self-fertile, almonds yield significantly better crops if one or several additional trees are planted nearby. The pollen from peach trees will also pollinate almond flowers because of the two fruits' close biological relationship.[7] Make sure to select cultivars that are compatible, because many almond varietals are listed as having specific named cultivars that are best suited to match the time of season when each tree's flowers open.[8] Almond blossoms are reliant on bees and other early flying insects for pollination and are among the first flowering trees to open, when the number of pollinators may be limited.

Site and Soil Conditions: Almonds are fairly adaptable trees but must be planted in well-drained soil with a full sun exposure or they will decline over time. Peach rootstocks are commonly used for almonds because they are adaptable to a wide range of soil conditions. A sheltered location that is not exposed to heavy winds or frequent late frosts will also help prevent almond flowers from getting damaged, since they are some of the earliest blossoms to open.

Hardiness: Many almond trees are rated to zone 5, or −20°F (−29°C), but in the colder portions of the northern United States, care should be taken to select varieties that have a long track record for surviving hard winters. The early blooming time for almonds is the single biggest factor determining their chance for survival, but some trees have been grown as far north as Connecticut and even Vermont in the United States.[9] Our two trees have been repeatedly exposed to below-zero temperatures with late frosts and have survived for seven growing seasons.

Fertilization and Growth: Almonds require high amounts of nitrogen and phosphorus, so the trees will benefit from a yearly layer of mulch with organic nutrients such as compost and aged manure to improve the productivity and the health of the trees.

Cultivars: There are over forty varieties of grafted almonds selected for high-quality nuts, and these are generally available through specialty mail-order nursery catalogs.[10] All of the cultivars listed below have a reputation for productivity, hardiness, and yielding high-quality nuts.

'All-in-One': A late-blooming, self-fertile tree that produces high-quality soft-shell nuts with a sweet flavor.

'Bounty': A late-blooming self-fertile cultivar that produces large sweet almonds.

'Hall's Hardy': One the most hardy and late-flowering of almonds, this tree is self-fertile and produces bittersweet nuts that should be cooked before eating.

'Oracle Almond': A late-blooming and partially self-fertile variety that produces high-quality almonds, developed in Ukraine by the Nikita Botanic Gardens in Kiev.

Related Species: As stated in the introduction, the common peach (*P. persica*) is one of the closest relatives to the almond. *P. davidiana*, a Chinese wild flowering peach that produces inedible fruit, is also closely related to the almond.

Propagation: Almonds can be successfully grafted onto peach or bitter almond rootstocks. They can also be grown from fresh seeds by removing the shells after harvesting, then planting them in soil that is exposed to cold for two to three months (cold stratification).

Pests and Problems: Almonds are part of the *Prunus* genus, so they may be subjected to many of the same pests that affect stone fruits like peaches and cherries, such as brown rot, black knot, leaf curl, and mold. These problems will tend to get worse if the growing season is particularly rainy or the soil stays soggy for long periods of time.

American Chestnut

Castanea dentata

American Chestnut 'Dunstan'

All words about the American chestnut are now but an elegy for it. This once mighty tree, one of the grandest features of our sylva, has gone down like a slaughtered army before a foreign fungus disease, the Chestnut blight.

—Donald Culross Peattie, 1950[1]

The American chestnut was one of the great trees of American forests that provided delicious edible nuts for wildlife and people as well as quality hardwood for carpentry. This magnificent tree was distributed across large portions of eastern North America but was wiped out in the span of fifty years. The culprit was an invasive chestnut blight (*Cryphonectria parasitica*), which was discovered in New York State in 1904 and quickly spread across eastern North America, killing nearly four billion American chestnut trees. The blight reached the United States through the introduction of Japanese chestnut trees (*Castanea crenata*), that were imported into the United States as early as 1876.[2]

Fortunately, many scientists, botanists, and forestry specialists love native chestnut trees and have been developing blight-resistant cultivars for the last sixty years in order to save this indigenous tree from extinction. In the 1950s an American chestnut tree that had survived the blight in Ohio was collected as scion wood for grafting and sent to the American plant breeder Dr. Robert T. Dunstan. Since Chinese chestnuts are immune from the blight, Dunstan cross-fertilized American chestnuts with their pollen. The work of ten years of his plant breeding resulted in the 'Dunstan' chestnut, a blight-resistant tree with a portion of the Chinese gene pool, and these trees have survived planting trials throughout North America.

Dr. Dunstan was not the only person interested in saving the American chestnut. Many organizations such as the USDA Forest Service, American Chestnut Foundation, and Connecticut Agricultural Experiment Station also carried out breeding programs for many years to produce blight-resistant American chestnuts with the goal of reestablishing native chestnut trees, and these efforts are still in the process of being tested.[3] Saving this wonderful tree from extinction is a work in progress because the improved, new cultivars have disease resistance but are not 100 percent free of the pathogen. These experiments and new efforts in the future may someday restore the glory of American chestnuts to our forests.

If you have room for a few large trees on your planting site, we believe that it is well worth growing blight-resistant cultivars in order to harvest these delicious nuts each season.

Growth Difficulty Rating: 1. Blight-resistant trees have performed well in sites all over the United States. Once blight-resistant trees get established, they are easy to grow.

Taste Profile and Uses: A yellowish green burr husk that is covered in numerous sharp spines splits open in the fall to reveal two to four shiny dark brown nuts that are ¾ to 1 inch (1.9–2.5 cm) in size. Chestnuts should be allowed to ripen on the tree and not shaken off branches, because they accumulate more than 50 percent of their volume in the final two weeks of ripening before dropping off the tree.[4] We have found that when the nut burrs first begin to crack open, they are ripe enough to be picked off the trees with leather gloves. If they are brought indoors and stored inside a cardboard box, the burrs will continue to crack open over a period of weeks out of the reach of squirrels.

Raw chestnuts have a bitter, acidic taste and can only be eaten after being cooked. They must be pierced before cooking and are

Male chestnut flowers.

The small female flowers.

The spiny chestnut burr husks.

Ripe chestnuts in spiky burr.

Chestnut foliage with male flowers about to open.

never consumed raw unless they are dried and ground into a flour. Boiling or roasting the nuts makes the meat inside sweet and soft with a rich starchy texture. The nuts are high in fiber and include minerals such as potassium, magnesium, and iron, with vitamins C and B_6. We think chestnuts are best eaten straight out of the oven, but these nutritious nuts can also be sprinkled in salads, puréed, baked in pastries, or added to desserts. Chestnuts ground down into flour are also commonly used as a substitute for traditional flour.

Plant Description: American chestnuts are the hardiest and largest tree in this genus, growing into a large 40- to 50-foot (12.2–15.2 m) tree, and in the past there were champions that were over 100 feet (30.5 m) tall. But more often the trees reach a smaller size in the form of a broad open canopy with attractive dark green foliage with prominent teeth along the edge of each leaf. The 6- to 10-inch (15.2–25.4 cm) leaves are alternately arranged along the branch and are shiny and waxy with a leathery appearance. The bark of young trees is smooth and olive to chestnut brown when young, then becomes gray and lined with deep interlacing furrowed ridges that run vertically on mature trees.

Flowers: The showy catkins (male flowers) are composed of small white blossoms produced on dangling clusters that are 4 to 7½ inches (10.2–19.1 cm) long. The female flowers can be produced in small groups or singly, and appear toward the bottom of the male flower spikes. Each floret has an ovary with six slender, white styles, sitting in an urn-shaped calyx, and supported by prickly bright green little scales.

Pollination Requirements: Two chestnut trees are necessary for the formation of nuts. The pollination is predominantly done by the wind, but insects also play a minor role in cross-pollination.

Site and Soil Conditions: Chestnuts are tolerant of rocky, acidic, and poor-quality soil but prefer moist, well-drained planting sites with lots of sun, although the trees will grow in part shade. Chestnuts do not grow well in soggy, wet soil conditions.

Hardiness: The trees are hardy to zone 5 or −20°F (−29°C).

Fertilization and Growth: Like most forest trees that get extra nutrients from dropped leaves each fall, American chestnuts would benefit from an annual application of leaf mold or organic compost. In our experience these trees have the capacity to yield nuts at an early age—we have seen 4-foot (1.2 m) trees produce flowers.

Cultivars: If you want to plant American chestnuts on your property, the most important factor will be blight resistance. Even if you are living in the parts of the United States that are currently not affected by the blight, this may be a problem in the future, so we recommend that you choose blight-resistant cultivars. At the time of writing this book, several nurseries were offering American chestnuts with various levels of blight immunity, but most of these trees have not been tested over a long period of time, with the exception of 'Dunstan' chestnuts, which have survived without a blight infection for over fifty years. Hopefully in the future there will be a wider selection of blight-resistant chestnut trees available in the nursery trade.

'Dunstan' Chestnut: A blight-resistant tree that produces high-quality flavorful nuts. We have two of these trees; one of them was only 6 feet (1.8 m) high and started producing nuts after it had been in the ground for only three growing seasons. 'Dunstan' chestnut is the only nut tree to receive US plant patents.[5]

Related Species: European chestnut (*Castanea sativa*) are large-sized trees that do not have blight resistance. The American chinquapin (*C. pumila*) is a 6- to 10-foot (1.8–3.1 m) tall shrub species of chestnut found across the southern parts of the United States. It produces small nuts and does not have blight resistance.

Propagation: Chestnut trees are propagated by grafting and are also readily started from seed after a period of cold stratification. It is best to pot up fresh chestnuts in soil and store them protected from hungry critters over the winter.

Pests and Problems: *Cryphonectria parasitica* is the blight fungus that killed American chestnut trees. The blight remains present aboveground but is not transmitted to the root systems of the trees, so many dying American chestnut root systems continue to send up sprouts. Then when the branches reach around 6 to 10 feet (1.8–3.1 m) high, they, too, die from the disease. Remember that disease-resistant is not the same as disease-proof. Because the blight may be in your environment and can affect new trees, it is advisable to remove older, dying chestnuts, including their root systems, from your planting area in order to reduce the risk of infecting your new blight-resistant trees.

Phytophthora cinnamomi is a root rot mold that affects some chestnut trees growing in wet soil conditions and is primarily found in the southern parts of the United States. This is best avoided by planting chestnut trees in well-drained soil.

American Persimmon

Diospyros virginiana

Common Persimmon, Possumwood

They have a plomb which they call pessemmins, like to a medler, in England, but of a deeper tawnie cullour; they grow on a most high tree. When they are not fully ripe, they are harsh and choakie, and furre in a man's mouth like allam, howbeit, being taken fully ripe, yt is a reasonable pleasant fruict, somewhat lushious. I have seene our people put them into their baked and sodden puddings . . .

—William Strachey, 1600s[1]

Diospyros, the genus name for persimmon, literally means "fruit of Zeus," but "fruit of the gods" is also commonly used to describe this fruit when perfectly ripe. The persimmon is part of the Ebony family, a large group containing over four hundred species that are mostly tropical and subtropical.[2] American persimmons grow as a wild species across wide portions of the United States from Connecticut across to Ohio, then south through the Midwest to Texas and Florida. Most people are familiar with the Asian persimmons (*D. kaki*),

Persimmon flowers.

'John Rick' persimmons.

A 'John Rick' persimmon tree.

The attractive, patterned bark.

since it is the fruit they may encounter in the fall at their local food market or grocery store.

Unfortunately the Asian trees are not reliably cold-hardy in many of the northern latitudes of the United States, because they cannot survive temperatures that drop much below 10°F (−12°C). Asian persimmons are large because they have been cultivated for hundreds of years and in their wild state are closer to the size of our native persimmons.[3]

Since ancient times, Indigenous American tribes used persimmons as a foraged food crop. Thus they remained a wild (and smaller-sized) fruit for a longer period of time. It wasn't until the late nineteenth century that American persimmon cultivars started being selected for improved edible qualities.[4]

American persimmons are not as sweet as their better-known Asian sister species and are usually a smaller size. That being said, they make up for the lack of sweetness by having a more complex, rich flavor that is absolutely delicious. Because American persimmons have few problems and produce wonderful, late-season fruit, we feel a grafted American persimmon cultivar that is self-fertile would be a worthwhile addition to a home fruit tree collection.

Growth Difficulty Rating: 2. Persimmons are sometimes difficult to plant because they have stiff roots that are easily damaged (see the "Site and Soil Conditions" section below). But once established, American persimmon trees are easy to grow, with few pest problems, and thrive in most environments that get consistent moisture.

Taste Profile and Uses: When ripe, American persimmons develop golden-orange to red-orange skin. The fruit is round and grows between ¾ and 2½ (1.9–6.4 cm) inches in size. The fruit usually contains seeds, but there are a few seedless cultivars. When fully ripe, persimmons become soft and squishy, sometimes developing a waxy bloom or black markings across the outer skin of the fruit.

The best American persimmon cultivars ripen long enough to lose their astringency and develop a mild butterscotch or molasses-like flavor. American persimmons must *ripen on the tree* in order to be eaten, because unripe fruits are astringent and horrible to eat. The most effective method to find dead-ripe fruit is to search the bottoms of trees each morning for the sweet ones that have fallen to the ground from the previous night. This happens in our gardens from late September through November, when the fruit develops a delightful pudding-like texture that is among the best and last fruits of the fall season. Our favorite way to eat persimmons is out of hand, but the fruits can also be frozen, made into sauces, and used as an ingredient in a persimmon pudding, which is a baked dessert bread.

Plant Description: Wild American persimmon trees can grow from 35 to 60 feet (10.7–18.3 m), but in rare instances reach up to 80 feet (24.4 m) in height. Thankfully, grafted cultivars tend to be smaller, growing from 15 to 30 feet (4.6–9.1 m) in height. Mature trees produce handsome dark gray bark, developing lined patterns of thick blocky plates running down the trunk. The overall effect is a handsome specimen tree with patterned bark and dark green, elongated ovate leaves that are pointed at the tip with a leathery texture, which grow to 2¼ to 5 inches (5.7–12.7 cm) in length.

Flowers: Small cream white to pale yellow bell-shaped flowers develop toward the ends of one-year-old branches and form in the axils of the new branch shoots in late spring. Male flowers tend to be produced in groups of two to three and are about ⅜ inch (1 cm) long, while female flowers tend to be solitary and are slightly larger than ½ inch (1.3 cm).[5]

Pollination Requirements: Persimmons have a complicated sex life, one that has evolved over time. Generally speaking, wild American persimmons are dioecious (trees having either all male or all female flowers). But there is a strain of persimmon trees from the upper midwestern forests that have larger fruit and are more cold-tolerant. Some of these trees also produce both male and female flowers on the same plant and are partially self-fertile. This group of trees has often been used to produce some of the modern, improved fruit cultivars. Still more confusing, there are even some persimmon trees that are parthenocarpic, which means they set seedless fruit without pollination.[6]

Site and Soil Conditions: These trees are best cultivated in full-sun growing conditions for fruit production, but American persimmon trees will survive in part shade. As a widely dispersed native tree, it is able to survive in a variety of growing conditions, such as soil that is dry, areas with heavy clay content, rocky hillsides, sandy soils, and infertile areas. Michael Dirr reports that persimmons even grew in southern Illinois on abandoned coal-stripped lands.[7]

Persimmons have brittle black-colored roots that break easily, so be careful when you are planting them. Once you've dug a hole, try gently tamping the soil around the roots, taking care not to move them around too much. Gently fill the soil in around the base of the tree, making sure that it is planted with soil that is firm to the root collar. Persimmons develop long taproots and do not transplant well.

Hardiness: Widely reported to be hardy to zone 5, or −20°F (−29°C), with many cultivars that will survive to −25°F (−32°C). The main problem with growing a persimmon at the northern end of its hardiness range is having a growing season long enough to allow the fruit to ripen on the trees. If you are planting persimmons in the coldest part of zone 5 or colder, try to select cultivars that have been tested as producing fruit in that zonal range.

Fertilization and Growth: Persimmon trees seem to thrive with neglect, but like all forest trees they benefit from an annual addition of leaf humus or organic compost.

Cultivars: All of the cultivars below are listed as self-fertile and will produce some fruit when planted alone. But fruit harvests will be increased if a pollinator or a male tree is planted with them. All of these trees are rated to be hardy to −25°F (−32°C).

'Early Golden': A nineteenth-century cultivar that is the mother of many of the modern persimmons and is one of the hardiest trees. Produces moderate-sized sweet fruit.

'John Rick': One of our favorite trees, consistently self-fertile. Produces fruit that is up to 2 to 2¼ inches (5.1–5.7 cm) in size, with a rich, sweet flavor.

'Szukis': A small-sized, slow-growing tree that is reliably self-fertile. Produces sweet medium-sized fruit that is seedless if planted alone. Will also pollinate other persimmon trees.

Fruit forming after pollination.

Related Species: Asian persimmon (*D. kaki*) is more widely sold in Asian and American food markets. It is winter hardy to zone 8 or about 10°F (−12°C), although there are a few cultivars of Asian persimmon rated hardy to 0°F (−18°C) or colder, such as 'Saijo' and 'Ichi-Ki-Kei-Jiro.' Inside the protection of an unheated greenhouse, we have been able to successfully grow an 'Ichi-Ki-Kei-Jiro' tree.

Asian and American Hybrids: 'Nikita's Gift' is a hybrid of Asian and American persimmon with fruit that is larger than American persimmon and is cold-hardy, rated to −10°F (−23°C). Our ten-year-old tree is planted in soil with a heavy clay content and does not perform well in growing seasons that are very wet.

Propagation: American persimmons can be raised from seeds taken from the fresh fruit, although the fruit will be of variable quality. Remember that persimmon trees are taproot-driven, and the seedling roots will often be the same length as the growth above the soil. The best-quality fruit comes off grafted varietals, so persimmons are traditionally grafted in spring in order to produce trees identical to the cultivars they were taken from. It is best to use ¼-inch (6 mm) thick hardwood cuttings selected for healthy large buds.

Pests and Problems: Our persimmons have only had a few minor pest problems, and a small amount of any of these will not affect the tree's fruit production, but they do deform the foliage. A black leaf spot is sometimes a minor problem. It is sometimes present on our tree's leaves, often spreading more during wet growing seasons. Fall webworm moths (*Hyphantria cunea*) sometimes lay eggs on our trees, and we have to remove a few caterpillar nests to stop their spread—again, this is more aesthetic than serious. Persimmon psylla (*Trioza diospyri*) occasionally attacks our trees. The small, white, powdery-looking nymphs deform foliage. In small amounts this kind of damage is not a major problem, and squashing the nymphs with fingers or removing infected leaves can limit the pests' spread on your trees if they show up in your garden.

Droughts and Root Suckering: While established persimmon trees are drought-tolerant, the foliage can get haggard and beaten-up looking during long, dry summers. Again, we find that this is normally an aesthetic concern and does not harm the fruit on our trees.

American persimmon root systems can form colonies of shoots that spout up around the trunk of the trees. If you have a grafted tree, the shoots should be carefully pruned to the ground or they may cause the tree to decline by taking nutrients from the original grafted cultivar.

Arctic Kiwi

Actinidia kolomikta

Chinese Gooseberry, Kiwiberry, Variegated Kiwi Vine

> *The foliage is purplish when young, and later in the season is more or less variegated, then sometimes at the apex, sometimes half the leaf, and occasionally the whole leaf being white or pink . . . The chief merit of this climber is in its curious and often very striking leaf-colouring.*
>
> —W. J. Bean, 1929[1]

Although kiwis belong to a family of vining plants with about forty different species that are distributed in tropical areas, with a few species native to temperate forests, most people have only encountered kiwi (*Actinidia delicosa*) in their supermarket produce section as a tropical fruit with bright green flesh surrounded by fuzzy brown inedible skin.[2] Arctic kiwi (*A. kolomikta*) is the lesser-known sister species to the common supermarket kiwi and is not a tropical native but found in temperate forests as a wild vine in Manchuria,

Korea, Japan, and northeastern China. As the name *arctic* suggests, the vines also survive in brutal winter environments like northern Russia, where the temperature drops to −40°F (−40°C). Russia has also had extensive breeding programs that introduced many of the best cultivars selected for larger-sized fruit, earlier ripening times, sweetness, and vitamin content.[3]

Although much smaller in size than the tropical species, the arctic kiwi has the distinct advantage over supermarket kiwis of ripening with a thin, edible skin that has no fuzz, so you can just pop them in your mouth when you want to eat the fruit. In addition to the culinary convenience of a thin, edible skin, another wonderful attribute of the arctic kiwi is that the male plants of this species produce foliage with magnificent variegated silver patterning. In fact, both male and female vines produce flushes of scarlet-pink that fade as the season progresses. Which is why these woody vines were originally introduced as ornamental garden plants, with little emphasis placed on their edible fruit.

Although kiwi was a traditional Asian wild food, Western culture was introduced to arctic kiwi by the great plant explorer Charles Maries, who collected seeds in 1878 for the famed Veitch and Sons seed nursery.[4] The English nursery that introduced arctic kiwi only lists it as an ornamental vine with no mention of the fact that the vines produce delicious fruit.[5]

By the 1990s in North America, the vines started to appear in garden literature and have been slowly introduced in the nursery trade and through mail-order fruit catalogs.[6] The plants are gaining attention throughout the United States and deserve to be better known because they are attractive and adaptable vines that produce good-quality fruit with no major pest problems.

Growth Difficulty Rating: 2. Kiwis are easy to grow, but because the vines can grow up to 20 feet (6.1 m) a season, they will require a strong support system such as a trellis to keep the fruit manageable. The long lateral growth will eventually require some pruning to keep the vines within the boundaries of most garden settings.

Taste Profile and Uses: Kiwi fruits are actually classified as a berry and taste the same as their larger tropical relative. Their green-colored flesh ranges in size from about ¾ to 1 inch (1.9–2.5 cm) in length, with some varieties selected for larger-sized berries. The fruits are also a healthy food with high amounts of antioxidants and are a good source of vitamin C.[7] As mentioned in the section above, the fruit has a thin skin and can be eaten out of hand, which is how we feel is the best way to enjoy the fruit, or by using them in desserts or salads.

Arctic kiwis mature on the vine over a long harvesting period lasting weeks. We have found that underripe fruit can be picked and placed in a bowl to ripen indoors if they are being taken by birds and chipmunks. Because select cultivars and vines planted in different sun exposures can produce edible fruit over a long growing season, we have been blessed by bountiful years with fruit ripening on different vines for over a month. If they have not been bruised or grown past ripeness, the kiwis can be stored in the refrigerator for several weeks, and freezing the fruit is a good way to preserve them for long-term storage if you have an abundant crop. We have tried making kiwi preserves, but after cooking the fruit turns to an ugly gray-brown color. Although it tastes okay, it's hard to eat a gray jam. When you dry them in a dehydrator, kiwis are reduced to the size of a raisin, with a grainy texture.

Plant Description: Arctic kiwi is a large, rambling deciduous woody vine that can expand up to 20 feet (6.1 m) a year, with plants recorded as growing as large as 50 feet (15.2 m), but male plants tend to grow at a slower rate than the female vines.[8] *Actinidia kolomikta* vines produce heart-shaped foliage that is a deep green color with uniform, serrated leaf edging. Male plants and some female cultivars tend to have leaves that are adorned with striking frosted white patterning often blending into rich deep pink colors. The attractive color displayed on male foliage varies from plant to plant, with some vines sporting magnificent patterns while others produce less noticeable color. These patterns fade away as the season progresses, and this process occurs more rapidly when vines are growing in full sun.

Plants as young as three years old can begin fruiting, and after several more seasons a single female plant can produce copious amount of fruit. Although arctic kiwi is less vigorous than many other species of kiwi vines (such as Chinese kiwi), it still requires a strong support structure such as a trellis, wall, or fence that can bear the weight of a large established fruiting vine.

Flowers: The plant's flowers are often hidden by the vine's lush foliage growth, but seen at a close vantage point, kiwi blossoms display small white petals that surround delicate white stamens (male pollen organs) that are arranged in a circle in male plants.

Female kiwi plants can be differentiated from their male counterparts by a group of stigmas (pollen receptacles) above a light green ovary. That little ovary (potential fruit) is the surest way to distinguish between males and females; plants are occasionally mixed up by nurseries and incorrectly labeled. Kiwi blossoms produce a lovely fragrance that fills the air with a pleasant floral perfume in early spring.

Pollination Requirements: Kiwi vines are dioecious with separate male and female flowers produced on individual vines, so at least one male and one female plant are required for fruit pollination. A single mature male kiwi plant can provide enough pollen to fertilize up to six separate female vines. The flowers are pollinated by several species of bees.

Site and Soil Conditions: We have found that arctic kiwi is an adaptable plant that prefers nutrient-rich soil that is well drained, but it will grow in average soil conditions. Some garden literature states that kiwis require large amounts of fertilization for good crops, yet we have seen that with an annual application of compost the vines in our gardens have been very productive.[9] In fact we have three 18-year-old vines under the shade of a butternut tree (*Juglans cinerea*) that have thrived and produced modest crops, despite the fact that butternut trees can adversely affect the plants that grow around them by producing a toxic compound (juglone) in the soil (see the "Juglone" sidebar on page 79).

Arctic kiwi vine is adaptable to a full-sun or part-shade location with the caveat that vines in full-sun locations will produce larger amounts of fruit. But plants located in a part-shade setting can still yield fruit because kiwi vines are native to Asian forest regions where they grow into tree canopies.[10]

Hardiness: Zone 3. Kiwi vines possess exceptional cold tolerance and are hardy to −40°F (−40°C), but many sources list these plants as being able to survive even colder temperatures. If you are gardening in the coldest parts of the United States, it is advisable to protect young kiwi plants—which have shallow root systems—for a few seasons of growth with a thick layer of mulch, but after that the vines are usually resilient enough to survive the harshest winter conditions. Flowers can be killed by a late-spring frost.

Fertilization and Growth: Kiwis are shallow-rooted vines from forest and scrubland areas, and like all plants that are native to the woodland environments they would benefit from a yearly application of leaf mulch, garden compost, or well-aged manure.

The colorful leaves of the male plant.

Arctic kiwi female flowers and young fruit.

Arctic kiwi leaves.

Several vines growing over a circular trellis.

Cultivars: Male plants are commonly sold in the nursery trade as "male" vines for pollination, and there are roughly about twenty female fruiting cultivars of arctic kiwi selected for larger-sized fruit and ornamental traits, with a few new ones being introduced each year.[11] Some of the female cultivars are rare and difficult to find, although a few noteworthy varieties are more common in cultivation. Here are a few good female varieties that are currently more widely available.

'Hero': A productive variety that produces large sweet-flavored fruit.
'September Sun': Although a female vine, this variety was selected for strong leaf patterns and good-quality fruit that ripens later than other varieties.
'Viktor': A vine discovered in the wild in Vladivostok, Russia, that produces distinctive lozenge-shaped fruit double the size of most wild fruit.

Related Species: Silver vine (*A. polygama*) is another interesting member of the kiwi family whose fruit tastes like a mixture of kiwi and mild chili pepper; it also produces attractive variegated foliage.

Propagation: Kiwi is commonly propagated with firm softwood cuttings about 1/4 inch (6 mm) thick that are taken in early summer, or by using hardwood cuttings potted up in soil in the early part of winter and grown out in a shady location for a couple of seasons. The easiest way to produce new plants is by layering stems in the soil for two growing seasons, until they have produced enough of a root system to be separated from the mother vine and replanted in another garden spot.

Pests and Problems: Arctic kiwis are not bothered by any major pest problems. The vines do not like waterlogged soils and may develop problems if planted in wet areas.[12] Squirrels and chipmunks like the fruit, and birds will eat the fruit as well and use the vines for nesting.

Asian Pear

Pyrus pyrifolia, P. × bretschneideri, P. sinensis, P. ussuriensis

Chinese Pear, Japanese Pear, Korean Pear, Sand Pear

There is a story current among the foreign residents in China that a certain newcomer was asked his opinion of the Chinese pears. "Well," he said, "it depends on what you eat them as; as turnips they are certainly fine, but as pears I would rather not express any opinion."

—FRANK N. MEYER, 1911[1]

That derisive comment about Asian pears was said to the great food and plant explorer Frank Meyer regarding the "culinary value" of the fruit cultivated in China in 1912. But Meyer was smart enough not to be dissuaded by the opinions of foreign travelers that were more familiar with European pears (*Pyrus communis*) and had only sampled a few Asian pears in a giant country that probably had over two thousand varieties. While some of the Asian pears that Meyer tasted were bland with a mealy, rough texture, he had already collected

The striking Asian pear blossoms.

Young fruit developing on the branch.

'Kosui' fruit.

'Yongi' fruits.

over forty distinct Asian pear cultivars, and many of the fruits he had sampled had, "hard flesh, but [were] very juicy and sweet."[2]

When going through this chapter of the book, bear in mind that most of the Asian pears in the modern world are derived from the sand pear (*P. pyrifolia*),[3] a tree native to China and Japan that is often crossed with other Asian wild pear species such as *P. ussuriensis* and *P.* × *bretschneider*. These trees have been cultivated for so long (possibly three thousand years) that the gene pool has been intermixed, and it is difficult to tell exactly which species is the Asian pear.[4] The old name *sand pear* was bestowed upon the fruit because the wild species of the trees produce fruit with a dense, gritty texture "like sand" that made them unappetizing. Over hundreds of years Asian horticulturists developed thousands of cultivars of this fruit, which is now one of the great fruiting trees of the world with a large range of skin colors, fruit sizes, and flavors. After having doubts about the quality of Asian pears, we planted a tree with three different varieties grafted onto a single rootstock, and it produces some of the sweetest pears that we have ever tasted.

The Japanese name for the Asian pear is *nashi*, and Japanese fruit breeders have probably done the most to breed and elevate the quality of the Asian pear in the modern world. These pears were not widely available in American food markets until about a decade ago. Even now the fruit choices are very limited considering this fruit has thousands of varieties.

A few good cultivars of Asian pear would make a great addition to any garden that has enough space for two trees. One of the best reasons to grow this plant is that it produces fruit at a young age while its close European pear relatives commonly take up to ten years. Asian pears are also one of the most consistently precocious fruiting trees, and we have seen very young specimens that were only 5 feet (1.5 m) tall yield small crops the year after being planted.

Growth Difficulty Rating: 1. Asian pears are easy to grow with a few pest problems that affect all pear trees, although these are less susceptible to those problems than their European relatives.

Taste Profile and Uses: The Asian pear has a wide range of sizes and colors with a shape that is usually round (with some exceptions) and slightly flattened on the top and bottom. The fruit is technically considered a pome, which has a thickened outer fleshy layer surrounding a central core, commonly containing five seeds. Asian pears cultivars can vary in size from 2 to 3 inches (5.1–7.6 cm) wide, but several cultivars produce fruit 4 inches (10.2 cm) across. Like every plant with thousands of varieties, the Asian pear's fruit comes in a range of colors including yellow-green, pale cream-yellow, deep russet brown, and olive-brown with fruit surfaces that are an even color or overlaid with attractive speckled patterns. The taste of the Asian pears can also depend on the variety selected. Some are sweet with a luscious, complex flavor that combines pear and butterscotch; other types are mild with a crisp, crunchy texture; and still others firm and blander tasting. One negative aspect of some of these cultivars is that the flesh of the fruit contains a gritty texture known as stone cells, and eliminating this particular trait has been one of the main focuses of plant breeders. In some years a single tree can yield fruit with a range of quality from firm and delicious to mealy; this may be the result of insect pests, environmental factors, or viruses.

Unlike European pears, which must be picked firm and ripened off the tree, Asian pears must ripen on the tree. After picking, the flavor of the fruit will be at its best for about two weeks, or they can be chilled in the refrigerator for months. Handle them gently—the fruits bruise easily. The fruit is best eaten out of hand, added to salads, or used in both sweet and savory recipes in traditional Korean and Japanese cuisines.

Plant Description: The 2- to 4-inch (5.1–10.2 cm) long leaves are broadly ovate, with sharply serrated edges that are glossy green. Since Asian pears are a combination of different wild species, there is a great diversity in the range of heights that a mature tree can grow, from 10 to 30 feet (3.1–9.1 m), with a wide spreading crown. Many modern varieties that originated in Asia are grafted on dwarf pear rootstocks, which tend to produce trees that have a shorter stature and can be maintained through occasional pruning to about 8 to 20 feet (2.4–6.1 m). There are examples of wild Asian pear trees (*P. pyrifolia*) in cultivation that are one hundred to three hundred years old that have grown 50 feet (15.2 m) or more in height, but this stature is more common in wild seedlings and not in grafted cultivars.[5]

Flowers: Asian pear trees produce prolific amounts of attractive 1½-inch (3.8 cm) white flowers that open in late April through early May on fruit spurs (small, elongated growths that are 1 inch / 2.5 cm or less) on branches that are one to three years old. The flowers are produced in clusters of six to eight blossoms. This is one of the great decorative attributes of Asian pear trees and it improves as the trees age, with old specimens being completely covered with exquisite blooms.

Pollination Requirements: The number of wild species and cultivated trees that have been combined for breeding varieties has resulted in a few cultivars of the Asian pear trees being partially self-fertile, while a majority require another tree for cross-pollination. So it is prudent to plant two trees together for bountiful harvests. There are also late- and early-blooming Asian pear varieties with flowers that open at slightly different times of the season, so the best chance for getting consistent crops each year is achieved by planting the most compatible cultivars together. Grafted varieties sold through reputable nurseries usually list good pollination choices for the pear trees they sell. Asian pears bear flowers earlier in the spring than European pears, but it is possible to cross-pollinate Asian species with the early-blooming European pear varieties if you want both trees on your planting site.

Site and Soil Conditions: The trees require a full-sun planting site for good fruit development and to reach their full potential. Well-drained soils are best for Asian pear trees, but many of the rootstocks used for modern varieties are adaptable to soil having a heavy clay content, as long as the roots do not constantly sit in soggy conditions. If your growing area is a little wet, then planting the tree in a wide spreading mound may help the root system shed some of the moisture, but this will ultimately not help if the soil is constantly wet.

Hardiness: Zone 5. Many cultivars are cold-hardy to −20°F (−29°C), but some varieties can take temperatures of −25°F (−32°C).

Fertilization and Growth: Pears require a good amount of nutrients to produce large crops of well-formed fruit, so a yearly application of 2 inches (5.1 cm) of organic garden compost or leaf mulch around the radius of the tree trunk will improve your fruit yield. However, avoid using manure and fertilizers that contain heavy amounts of nitrogen because of the potential for spreading fire blight disease (see the "Pests and Problems" section). These trees can be long-lived and tend to be precocious bearers, so thinning the tiny fruitlets is important, not only to increase the size of the pears but also to reduce the risk of the limbs breaking from large loads of fruit. This can be accomplished by removing the small fruitlets after pollination and leaving 3 to 6 inches (7.6–15.2 cm) of space between fruits so that they have room to grow.

Cultivars: Although there are thousands of known varieties in Asia, there are far fewer grafted Asian pear cultivars available in the United States. About forty to sixty varieties have been cultivated in North America at various times, and nurseries may offer around a half dozen choices. Mail-order nurseries sometimes offer a small number of grafted fruit trees.[6] Below are a few generally available cultivars.

'Chojuro': A very productive Japanese cultivar from the late nineteenth century that is widely available. It produces good-tasting fruit that is mildly sweet.

'Korean Giant': A late-ripening variety famous for producing softball-sized fruits that have a sweet flavor. Best planted in areas with long, hot summers to fully ripen the fruit. Resistant to fire blight.

'Kosui': A Japanese variety that is a vigorous-growing tree, which yields moderate crops of extremely sweet golden-brown fruit with a crisp, crunchy texture.

Multigraft Trees: Some specialty fruit nurseries graft different types of Asian pears onto a single rootstock and offer them for sale as self-fertile, multigrafted trees with names such as three-in-one Asian pear or combo Asian pear. This may be a good choice for a garden that has enough space for only a single tree.

Related Species: The European pear (*P. communis*) is related to all of the wild pear species in Asia, as are all the members of the apple genus (*Malus*).

Propagation: The common method of propagating Asian pears is by grafting the scions onto a rootstock that is tolerant of different soil conditions and resistant to fire blight, such as *P. betulifolia*, but seedlings of other wild Asian pear trees are also used for rootstock.

Pests and Problems: Like their pear and apple relatives, Asian pears can suffer from bacterial problems like fire blight (*Erwinia amylovora*), which is caused by environmental conditions in areas that have large amounts of rain and humidity during the spring. These problems are often brought on by the rapid growth induced by fertilizers with heavy amounts of nitrogen, such as fresh manures.[7] Try choosing varieties that are resistant to this deadly bacteria. Codling moths can affect Asian pears and are considered a serious pest because the larvae burrow through the skin and calyx of the fruit, damaging it. Good hygiene practices such as clearing and placing damaged fruit in the garbage will help to stop the spread of these moths.

Beach Plum

Prunus maritima

Seaside Plum

I've known the beach plum since childhood on Cape Cod, where it was the only woody plant in the sea of dune grass that separated the ocean from the rest of the world.

—Richard H. Uva, 2003[1]

Beach plums are a scruffy wild fruit of the sandy regions that straddle the eastern coastline from North Carolina to Maine.[2] They're one of those tough, hard-to-kill plants whose wild fruit epitomizes the late-summer season in places like Massachusetts. The history of the beach plum as a wild food goes back to the Indigenous inhabitants of the United States. It was one of the first fruits used by European settlers to North America. It was included in the Pilgrims' first Thanksgiving and eaten by Henry Hudson in 1609.[3]

As an endemic wild fruit, beach plum has resisted the breeding efforts of American horticulturists who attempted to create an American beach plum industry. In 1875 Luther Burbank crossed beach plums with Asian plums and produced a large-sized hybrid called

'Giant Maritima'. Unfortunately these fruits were too soft to ship, so they never became commercially successful. Then in the 1930s agricultural groups in Cape Cod tried to start a commercial beach plum industry to produce canned jelly, but this project was abandoned when World War II broke out and the funding was diverted to other established commercial crops.

In the battle for shelf space in modern supermarkets, beach plums could never compete with Asian plums, which are larger and have thicker outer skins making them perfect for shipping.[4] Beach plums are highly variable in size, color, and flavor, and the bushes cannot always be relied upon to consistently yield fruit. So beach plums were relegated to being roadside wild plants that got overlooked in favor of more domesticated agricultural commodities. But for those who love walking through an open beach site so that they can get a taste of native American fruit, beach plums are well worth the effort. For difficult growing areas that would kill most fruit trees, a border of beach plums could provide lots of fruit to make jelly. As an edible privacy hedge, planted between houses in a suburban environment, beach plums could provide magnificent spring blossoms and summer plums that grow at good picking height for children.

Growth Difficulty Rating: 1. Beach plums are easy to grow once established, and thrive with little care.

Taste Profile and Uses: In all aspects of its habit and appearance, beach plums are wild and variable. Fruit size ranges from about ½ to 1 inch (1.3–2.5 cm), with a substantial inner pit in relation to the fruit's actual size. In the wild the color of beach plums can be dark blue, purple, crimson, or yellow in rare instances; the skins often have a heavy bloom. For the best flavor, they need to ripen on the plant, which typically occurs mid-August through early September. The flavor of the fruits is variable, with some bushes producing delicious high-quality sweet-tasting plums, while others produce sour fruit better suited for making jelly. Our bushes that produce larger-sized fruit are usually the sweetest. At its best, beach plum takes on the distinct taste of plum combined with berry, with hints of Concord grape. There are also isolated populations of yellow-skinned beach plums (*Prunus maritima* var. *flava*) that are reported to be one of the best-tasting forms of this fruit.[5]

There is a long history of New Englanders using beach plums as a wild-harvested fruit for making jelly, fruit sauces, and

Beach plum hedge in spring.

Beach plum flowers.

Beach plums developing.

A branch full of plums.

marinades for meat barbecuing. The bushes' fruiting habit can be unreliable, and the quantity of fruit varies from year to year. There are times when the plants produce bountiful harvests, but in other years they yield smaller amounts. We feel that the smaller yields may be the result of heavy, prolonged rain during their flowering period, resulting in a loss of the pollen.

Plant Description: The dull green leaves are alternately arranged along the stem and grow 1½ to 3 inches (3.8–7.6 cm) long with lightly serrated edges. Beach plum is a multi-stemmed deciduous shrub spreading by root suckers, often forming dense thickets that grow to around 6 feet (1.8 m) high and wide—though under cultivation the shrubs can grow taller and occasionally reach 10 feet (3.1 m). Bushes can also be pruned into single-trunked trees that reach about 10 to 15 feet (3.1–4.6 m) high.

Flowers: Small, five-petaled, fragrant white flowers bloom in the early spring before the leaves open. A group of beach plums in bloom is a magnificent sight, with huge flower displays that cover entire bushes. In 1932 the Arnold Arboretum staff were able to see beach plums blooming and plot their wild geographic distribution from the vantage point of an airplane.[6]

Pollination Requirements: For fruit-set, at least two bushes are required for pollination.

Site and Soil Conditions: *Prunus maritima* is native to beach areas, so the shrubs survive some of the worst growing conditions, such as heavy winds, high levels of salt, sandstorms, and poor-quality soil that would kill most fruiting plants. However beach plums can be grown without a beach, in most average soil conditions, as long as the roots have good drainage. Our bushes have thrived for a decade in soil with heavy clay content that is well drained. But the most important factor for a healthy plant is for it to be sited in full sun.

Hardiness: These are tough and resilient plants that are found growing in the wild in windy coastal areas up to Maine, to zone 4, or hardy to −30°F (−34°C).

Fertilization and Growth: Despite the beach plum's ability to thrive with neglect in harsh conditions, research has shown that plant vigor and fruit production are increased by fertilization, and a yearly addition of compost will help to create vigorous, healthy bushes.[7]

Cultivars: Beach plums have resisted fruit breeding efforts, but a few named cultivars have been created that may be available as grafted fruiting bushes. Because it is viewed as a wild plant with fruit that is smaller than commercial plums, it is often difficult to find beach plum cultivars in modern nursery commerce.

'Jersey': A productive cultivated variety with slightly larger plums that produces a 10- to 12-foot (3.1–3.7 m) tree.
'Premier': A cultivar that originated on Plum Island, Massachusetts, selected for its large 1-inch (2.5 cm) blue fruits.

Related Species: There are at least thirteen species of wild North American plums that produce edible fruits with a wide range of flavors.[8] *Prunus angustifolia* (Chickasaw plum) is an attractive suckering wild bush plum found in the southern United States that is hardy to zone 5, or to −20°F (−29°C). *P. nigra* (Canadian black plum) is a small-sized tree native to the Great Lakes region of the United States and Canada that produces sweet plums and is hardy to zone 3, or to −40°F (−40°C).

Propagation: Seeds are the easiest way to make more beach plum bushes. So when you eat particularly delicious fruit, save the pits because these are more likely to produce a plant that yields sweeter fruit. The seeds should be planted in pots with soil and protected from critters through the winter, and these should germinate after a winter dormancy period. Another way to produce more plants is through softwood and hardwood cuttings taken from mid- to late July.

Pests and Problems: Being a member of the *Prunus* genus, beach plums are subject to black knot (*Apiosporina morbosa*) and brown rot (*Monilinia* spp.), but these problems seem to occur in smaller amounts than they do on cultivated plum trees. Our unsprayed beach plums ripen over a period of two to three weeks; brown rot does affect our fruit but usually occurs in limited amounts and is present toward the end of the ripening season so that the majority of our harvest is not impacted by this fungus. Discarding the rotting fruit and pruning away black knot will keep these diseases in check.

Our biggest problem with harvesting fruit is wildlife. The bushes can be completely stripped of fruit by birds, squirrels, and chipmunks. So in the years when bushes produce a bountiful harvest, protecting them with a growing cover may enable you to eat more fruit, unless of course you want to share it with wildlife.

Black Raspberry

Rubus occidentalis

Blackcaps

> *This species is usually to be found in fence-rows, in copses, and along roadsides, a common and useful food-plant, although sometimes a pestiferous weed in vast regions throughout the extensive territory in which it is found.*
>
> —U. P. Hedrick, 1921[1]

U. P. Hedrick's comment illustrates America's mixed feelings about the value of the black raspberry as a cultivated crop that is a wonderful wild fruit for a wide range of uses but can often do too well in cultivation by growing aggressively and self-seeding baby plants until it overtakes other cultivated areas and becomes a weedy, thorny nuisance. *Rubus occidentalis* is a common denizen of the northern woods of the United States that is distributed from New Brunswick and southern Quebec down to Georgia and Missouri, and westward to Oregon, Washington, and British Columbia. For most of human history the wild black raspberry was an undomesticated crop eaten by Indigenous tribes and early European settlers. It was not until about

1832 that the first attempt at domestication of the wild berry succeeded in producing an improved cultivar called 'Ohio Everbearing' that was selected from a promising garden plant in Cincinnati.[2] From there the fruit gained some popularity as other cultivars were selected from promising wild plants with names such as 'Doolittle' (discovered in New York in 1850).

In an earlier era, black raspberries slowly began to be cultivated for use as market fruits that were more widely sold than the common red raspberry (*R. idaeus*), and by 1925 there were almost two hundred different cultivars listed for black raspberry.[3] But it was the black raspberry's susceptibility to disease under agricultural cultivation that ultimately proved to be the major strike against this wonderful fruit, which resulted in farmers abandoning it in favor of red raspberries, which are not as affected by fungal and virus problems.[4] Though modern fruit breeders have produced many cultivars that have better disease resistance, blackcaps have never fully recovered their popularity as a market fruit.

For many years our gardens had a wonderful stand of wild black raspberries that declined over time as the berries were hit by a fatal bacterial disease called orange rust that causes the berry canes to slowly decline and die. We began to miss the flavor of this delicious annual fruit and started a new berry patch a safe distance away from the diseased blackcaps by growing the disease-resistant cultivated variety called 'Allen', and within a few years this berry patch started producing huge quantities of tasty berries. While cultivated blackcaps may not possess a flavor as intense as the wild fruit, it is very close, and the fruits are absolutely worth growing.

Growth Difficulty Rating: 2. Black raspberry canes are easy to grow and among the most productive fruiting plants, often producing berries a year after planting. But eventually brambles will require some management to deal with the rampant growth of the canes. They ultimately need to be pruned to facilitate easy berry harvesting.

Taste Profile and Uses: The fruit is not an actual berry but a cluster of druplets or flesh-covered seeds clustered around a receptacle. Unlike a blackberry the torus (receptacle or core) is left behind when the fruit is picked. The berry size is about ½ inch (1.3 cm) round, but several bred varieties produce larger fruit.

Blackcaps possess a wonderful flavor that embodies the sweetness of a good red raspberry enriched by the sharp notes of a great blackberry. The berries are wonderful to eat fresh off the cane and can be used to make incredible desserts, smoothies, and jam. They can also be frozen and thawed out for later use. Given the diversity of what can be grown in North American gardens, we still strongly believe that black raspberry jam is the best of all preserved fruit. The chief drawback to the berry, and one of the main reasons it has not been as popular as the red raspberry, is a tendency for the fruit to produce larger quantities of seeds to a smaller proportion of fruit pulp than the red-colored sister species.[5] When making blackcap fruit preserves, this problem is easy to alleviate by passing the fruit pulp through a metal strainer in order to remove a portion of the seeds. The berries are also filled with a wide range of nutrients and antioxidants, such as anthocyanins and phytochemicals, which are used by the body to help fight cancer and other adverse health problems.[6]

Plant Description: A black raspberry bramble is composed of a series of shoots that sprout from the root system as canes (large arching stems), which grow between 3 and 12 feet (0.9–3.7 m) in length. The stems produce small curved thorns about ¼ inch (6 mm) long that grow on canes, which start out light green then mature into a nice purple color that is often overlaid with a glaucous white bloom. The canes grow coarse, dark green compound leaves that are 2 to 4 inches (5.1–10.2 cm) long with serrated edges, arranged in groups of three or five leaflets that have a distinct silvery white underside.

Life Cycle of Berry Canes: All berry bramble stems are biennial. In the first year a primocane (new cane) grows from a bud in the roots into a long slender, curving stem with leaves but no flowers. In the second year the original cane, now called a floricane, develops a second flush of leaves with flowers that ultimately produce the fruit. After fruiting the floricanes will die at the end of the growing season. A cane system often thrives for eight to ten growing seasons, and after this length of time the root systems begin to decline for several years until finally dying off. The longevity of berry brambles can be lengthened with a yearly addition of well-aged compost or leaf mulch to the soil surrounding the roots. Our patch has been productive for fourteen years.

Flowers: Black raspberry canes produce small blossoms in clusters of five to fifteen flowers that are up to ½ inch (1.3 cm) wide. Each flower has five white petals above five long, greenish white sepals; these form around the center of the flower, which is filled with a large collection of stamens and pistils. Each of the flower's pistils that gets pollinated can produce a tiny separate seed covered by juicy pulp, and all of those grow together to form a single edible fruit or drupe.

Pollination Requirements: The cane systems are self-fertile and often very productive, but because of the relatively small size of the fruit, it is generally necessary to plant several berry bushes together to increase fertility and to produce enough berries to really enjoy this quintessential summer fruit.

Site and Soil Conditions: Blackcaps grow in a wide range of soils but flourish to their greatest potential in nutrient-rich soil that is consistently moist and deep.[7] They are found growing in forest environments and will produce fruit in partly shaded conditions such as four to six hours of direct sunlight, although a growing area that is in full sun exposure will be the most productive. Since berry canes produce shallow root systems close to the soil surface, they do not like competition from weeds and grasses. Soil that is constantly wet and soggy can seriously affect plants and make them more prone to pests that will ultimately destroy most of their root systems.[8] Planting in a slightly wet area can be managed by growing berry canes on long rows of soil raised on a mound 6 to 8 inches (15.2–20.3 cm) above the ground. However, this may ultimately fail if the planting site is consistently too wet.

Managing a Black Raspberry Patch: There are dozens of considerations and ways to deal with berry canes, ranging from allowing the plants to form a "wild" bramble patch to formally training the canes onto a trellis system. We've listed several different strategies for managing berry brambles under "Managing a Blackberry Patch" on page 84.

Hardiness: Blackcaps are often listed as zone 5 or to −20°F (−29°C), but there is a range of winter temperature tolerance in cultivars. Many varieties such as 'Allen' and 'Jewel' exhibit greater cold resiliency, sometimes listed as hardy to zone 4, and several varieties are more sensitive to colder temperatures and are rated as zone 6.

Fertilization and Growth: Berry canes will grow with neglect, but applying a layer of garden compost, well-aged manure, or leaf mulch each spring will help the roots stay cool, and result in larger berry harvests.

Cultivars: In the wild blackcap, fruits are smaller but often have a richer flavor than the bred varieties, so look for disease-resistant cultivars that also have a good reputation for flavorful berries. Below

Black raspberry flowers.

Black raspberries on a trellis.

Black raspberry fruit.

we have listed a few that are commonly available and have a good reputation for disease resistance, cold hardiness, and rich flavor.

'Allen': A large-sized berry variety with a strong flavor that is consistently productive and very resistant to disease that was introduced by the Geneva (New York) Experiment Station in 1963. The fruit harvest can extend to three weeks in our gardens if it is not too rainy.

'Jewel': Another introduction from the Geneva Experiment Station. This is considered the most popular variety, producing substantial yields of large-sized fruit with good disease resistance.

'Munger': A cultivar from Ohio introduced in 1898 with a good reputation for producing large-sized flavorful fruit with a strong resistance to diseases.

Related Species: There is another black raspberry (*R. leucodermis*) native to western North America from Alaska to the bottom of California, as well as the many species of wild raspberry native to the United States.

Propagation: Blackcaps are commonly propagated by tip layering—burying the tip end of the canes in soil and allowing new roots to form. After a growing season they can be separated from the mother plant in the fall and potted up with soil for future use.

Pests and Problems: Black raspberries are fairly easy plants to grow, but starting out with disease-resistant varieties that are sited in open light and planted in consistently moist and well-drained soil will alleviate many of the problems that affect blackcaps.[9] If you have wild red or black raspberries on your property, try to locate your new plants a significant distance away from them, because viruses like orange rust (*Arthuriomyces peckianus*) can be passed from wild plants to new plantings. The biggest virus problems that affect the canes are anthracnose (*Elsinoë veneta*), mosaic virus, and verticillium wilt (*Verticillium albo-atrum*). When berries ripen on the canes, they can also be attacked by wasps, Japanese beetles, and some bird species. Gray mold can also be a problem on ripening fruit; it's brought about by prolonged exposure to rainy weather and can result in heavy crop losses if this occurs just prior to the fruit harvest.

Black Walnut

Juglans nigra

American Walnut, Eastern Black Walnut

In a more innocent age nutting parties were the most highly prized of children's festivities in autumn, throughout the eastern forest belt, and though butternut and hickorynut, hazelnut, chestnut and chinquapin, and even beechnut and kingnut were gathered, walnut was the favorite.

—Donald Culross Peattie, 1948[1]

Up until recently the American black walnut (*Juglans nigra*) was considered one of the great hardwood trees of the eastern forest, viewed as a handsome ornamental tree that produced prodigious crops of flavorful nuts and provided some of the finest decorative hardwood of all American trees. Indeed, when *American Forestry* magazine conducted a national survey in the early twentieth century and asked their readers what should be named the National Tree of the United States, a large portion of their audience voted for the black walnut.[2] These trees grow in mixed hardwood forests from

Male flowers.

Female flowers.

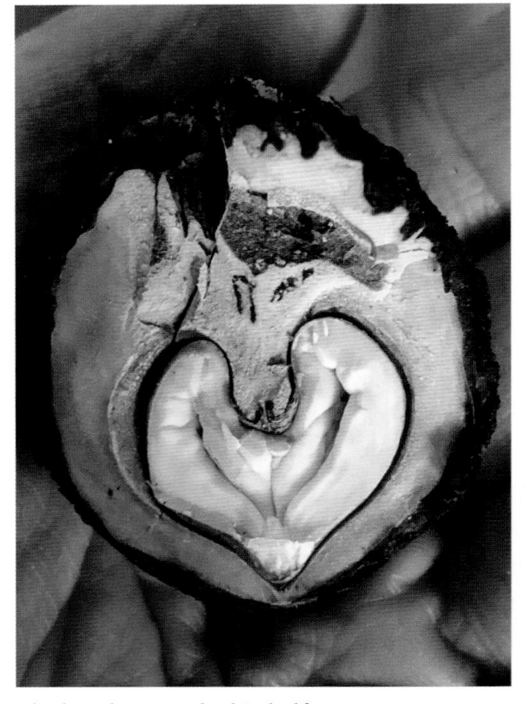

Black walnut cracked in half.

The handsome bark.

Massachusetts down to Florida and west to Minnesota and Texas, and the wood is still valued as one of the finest decorative hardwoods.[3] Indigenous tribes used the nuts as a valuable food to make cakes, flour, and butter, and tapped the trees for their sap.[4] Early colonial settlers from New England recognized the black walnut as a close relative to the English walnut of their native lands and used the nuts for so many food and household items that we would not be able to fit all of them on this page, but here are a few: Walnuts were gathered as a food crop and were used as a feed source for pigs and chickens; unripe nuts were pickled; and because the wood is strong and produces the most beautiful cut-wood planks in America, it was used in almost every type of household furniture that required a decorative surface, which is borne out by the fact that nearly all the baby cradles produced during the colonial period are made of this wood.[5]

Today this important native tree has fallen into disrepute; black walnut trees are viewed as a messy nuisance that litters properties with unwanted nuts. But if a few trees are located in the right place, they could literally provide an entire family (or several families) with hundreds of pounds of useful, delicious nuts a year for an entire lifetime.

Growth Difficulty Rating: 1. The trees are easy to grow with only a few minor pest problems.

Taste Profile and Uses: The nuts form inside a circular-shaped green husk that is 1½ to 2 inches (3.8–5.1 cm) wide that turns black when ripe. The taste of black walnuts is distinct, and some may find the flavor of the fresh nuts too strong and "green" tasting compared with the more neutral and blander-tasting nuts sold in most food markets, which are usually English walnuts. But many people feel that the wild native nuts are superior to commercial walnuts and their robust aromatic flavor is not diminished by cooking.[6] They are highly nutritious, and the nutmeats are oily with a strong, and even grassy, flavor that is a perfect addition to pesto, salads, baked goods, or black walnut ice cream. The husks have strong antibacterial components including juglone and iodine, and have been used as an insect repellent and for homemade medical treatments for killing parasites, treating fungal infections, eliminating warts, and soothing eczema.[7]

Extracting the Nuts: Mature black walnut trees can grow tall and produce their nuts at an imposing height that is out of reach until they ripen and fall to the ground. The nuts present two harvesting problems: The first is the green husk that covers the shell, and the second is the thickness of the shell. The inside of the husk is a dense black layer that surrounds the nutshell and is a formidable oily mess that stains hands. Removing this layer is best accomplished by using thick rubber gloves with a wire brush to strip off the oily husks. Then the walnuts need to be dried, and laying the nuts out in bright warm sunshine on a tarp for a two to three days is necessary to thoroughly dry them. After this the shell-encased nuts can be stored indoors. They can stay fresh for up to a year or more and be cracked open as needed. Cracking the actual nutshell can be a challenge because these are very thick and tough. We have placed the nuts on a concrete floor and pounded them with a hammer to loosen up the shells, but we find cracking them in a sturdy table-mounted nutcracker works best. But we have also heard of people literally driving a car over nuts in plastic bags to dislodge the shells!

Plant Description: Black walnuts are one of the largest trees in the walnut family and can grow from 50 to 75 feet (15.2–22.9 m) tall with wide spreading crowns, but individual trees have grown over 100 feet (30.5 m) tall. A full-grown tree is an impressive sight, a stately ornamental shade tree with a large straight trunk covered with deeply furrowed bark. The tree's aromatic foliage is also handsome with 2- to 5-inch (5.1–12.7 cm) long, yellow-green, lance-shaped leaflets arranged in groups of fifteen to twenty-three that form on attractive stalks that are up to 2 feet (61 cm) long.

Flowers: Black walnut trees produce both male and female flowers that are a light green color. The catkins (male flowers) grow in groups of drooping clusters of delicate blossoms that are 2 to 4 inches (5.1–10.2 cm) long. The female blossoms are generally produced in groups of two to five on the end of the current season's branches, and appear as small, teardrop-shaped ovaries with two tiny, rabbit-ear-shaped stigmas (the part of the flower for receiving pollen) on top.

Pollination Requirements: The trees are monoecious (separate male and female flowers growing on the same tree) and are pollinated by the wind, which means two trees or more must be planted together to ensure good cross-pollination. Walnut pollen can be carried on the wind for great distances—some sources suggest that it will travel hundreds of feet or more.[8] We think that the most consistent pollination results from planting two trees about 25 to 60 feet (7.6–18.3m) apart. This will also provide enough space for the large trees to grow in a healthy manner.[9]

Site and Soil Conditions: Black walnuts like open sites in full sunlight; however, they may tolerate a small amount of shade. The tree is adaptable to different types of soil conditions, including areas with a heavy clay content and areas with poor sandy soil, but it does best in rich, well-drained soil that is deep and receives consistent moisture throughout the growing season.[10] They can tolerate areas that are slightly wet as long as they are not constantly standing in water.

The two main considerations that should go into planting black walnuts are that they will grow into large trees, and that they may have negative effects on other plants growing around them. The latter is because of their capacity to release juglone into the soil (see the sidebar below). These are not appropriate trees for a small yard or garden area, but would be a great choice for a large open field near the edge of a yard or farm area where they can grow to their full potential and their negative effects on other plants can be minimized.

Hardiness: Black walnuts are hardy to zone 4 or to −30°F (−34°C).

Fertilization and Growth: The trees grow best in rich soil conditions, so they would benefit from a yearly amendment of decomposed leaf mulch or compost.

Cultivars: The improvement of black walnut trees has focused on varieties that produce nuts with larger kernels that are easier to crack and have better disease resistance.[11] These are somewhat

Juglone

All types of walnut trees produce a strong chemical compound called juglone, which is considered to be allelopathic. Allelopathy happens when specific chemicals produced by certain plants inhibit the growth of other plants around them in order to give a growing advantage to their own seedlings. All parts of walnut trees contain this substance, so choosing an appropriate growing site away from valued food and ornamental plants is crucial in order to minimize the effects of the juglone on these plantings.

rare in the nursery trade and may have to be ordered from specialty fruit tree nurseries.

'Elmer Myers': A very productive cultivar from Ohio selected for thin-shelled nuts.
'Emma Kay': A productive, high-quality variety that produces medium- to large-sized nuts with thin shells that are easy to crack.
'Thomas Black': A Pennsylvania cultivar introduced in 1881 that is the most widely grown variety, selected for large-sized nuts with thin shells.

Related Species: There are about eighteen to twenty species of walnuts distributed in temperate and tropical areas with six species native to the United States.[12] The English or Persian walnut (*J. regia*) is the most commonly cultivated edible walnut.

Propagation: Walnut tree cultivars can be grafted onto wild walnut tree rootstock, or trees can be grown from freshly harvested nuts. The germination rate is sometimes only 50 percent, but you will have the best success in producing plants by using well-filled nuts that sink in a bowl of water.[13] After this they can be potted up inside the shells in soil and should germinate after cold stratification for three to four months of winter. Small trees can be dug from the wild, but remember that black walnut trees produce taproots, which means that after three of four years the long central root of the tree will be deep in the soil and difficult to dig up without damaging the tree.

Pests and Problems: Thousand cankers is a disease that is killing black walnut trees and was found in 2008 in many western states. It is unfortunately spreading eastward and has been found in five eastern states as of the writing of this book. Two culprits, a fungus (*Geosmithia morbida*) and the walnut twig beetle (*Pityophthorus juglandis*), can kill an infested tree within three years. Other, much lesser problems include husk maggot flies (*Rhagoletis completa*), which do not usually bore through the nut shell but sometimes live inside the thick husk surrounding the shell. These can often be eliminated by removing the husks and thoroughly drying the shell-covered nuts. Anthracnose can sometimes affect the foliage of black walnut trees, while codling moth larvae (*Cydia pomonella*) can sometimes attack the nuts.

Often, though, the biggest problem is tripping over the fallen nuts.

Blackberry

Rubus spp.

> *. . . their habit of turning black before they are ripe nearly always led to them being gathered and eaten while green. . . . Moreover, their culture, being little understood, led to frequent failures and unsatisfactory results, while their propensity to persist and spread, aided by their unmerciful thorns, conspired to render them a terror to many timid gardeners.*
>
> —FRED W. CARD, 1914[1]

Blackberries purchased from a supermarket are a perfect example of how an extraordinary fruit has been destroyed by modern food production. Bred, developed, and picked for modern shipping techniques and supermarket longevity, these wonderful wild fruits have been rendered bland, unsweet, and flavorless. It is unfortunate that many people only know them as a tasteless berry sadly packaged in plastic, having never had the opportunity to taste ripe blackberries off a wild bramble. The genus *Rubus* has a huge geographic distribution spread across every continent except for Antarctica.[2] Wild blackberries were (and still are) often viewed as a thorny invasive weed and nuisance. But by the late nineteenth century, named cultivars were selected for favorable attributes. Unfortunately most of these have fallen by the weedy wayside, and now most people only encounter one or two varieties that have been bred more for their firmness than for their flavor.[3] Fortunately there are still good numbers of improved fruit cultivars available in garden centers and mail-order nurseries, with the current emphasis on plant breeders developing thornless varieties with great taste.

Growth Difficulty Rating: 2. Easy to grow, but all mature berry cane systems will eventually become unruly. You may need to manage their growth in a garden or field setting in order to conveniently harvest fruit without being harassed by thorns.

Taste Profile and Uses: The fruit is technically not a berry but an aggregate of drupelets connected to a core. Each drupelet is separately formed from the multitude of ovaries that each flower contains. Blackberry commonly ripens in the middle of summer between late June and August. During that time the best fruit will often fall into your hand when you touch them.

In the most flavorful wild and cultivated berries, the taste is a potent combination of acidity, floral notes, and sweetness concentrated in a 1- to 2-inch (2.5–5.1 cm) fruit that is as messy as it is delicious.

Besides the widespread use for jelly, jam, pies, cobblers, and fruit sauces, there is a long history of the blackberry fruit, roots, and leaves being brewed as teas and for medicinal purposes by Western herbalists.[4]

Plant Description: The leaves are produced in palmately compound (radiating outward from a central stem) groups of three to five (rarely seven). They have a wide range of sizes, from 1½ to 6 inches (3.8–15.2 cm) long, that often have course, doubly serrated margins. The upper portion of the leaf is a medium dark green while the undersides are a lighter color. Blackberry brambles are composed of a root system that grows horizontally, sending out groups of individual canes that emerge from the ground as slender arching stems that develop thorns, unless it is a thornless cultivar. These canes can grow between 6 and 18 feet (1.8–5.5 m) in length depending on the cultivar or species. The collection of all these canes usually forms an unruly tangle of stems that is referred to as a bramble.

Life Cycle of Berry Canes: All blackberry bramble stems are biennial. In the first year, a primocane (new cane) grows from a bud in

The attractive blackberry flower.

Blackberry growing on a trellis.

Blackberries ripening.

Blackberry fruit.

the roots into a long, slender, curving stem with leaves but no flowers. In the second year, the original cane, now called a floricane, develops a second flush of leaves with flowers that ultimately produce the fruit. After fruiting the individual floricanes will die at the end of the growing season.

Cane plantings often thrive for eight to ten years. After this length of time the root systems begin to decline for several years until finally dying off. The longevity of berry brambles can be lengthened with a yearly addition of well-aged compost or leaf mulch to the soil surrounding the roots.

Flowers: The flowers range in size from ¾ to 1 inch (1.9–2.5 cm) or slightly larger, with handsome, small five-petaled whitish pink flowers produced on short racemes with up to twenty blossoms (but generally fewer) that open up in late spring to early summer.

Pollination Requirements: Blackberry brambles are self-fertile and feature flowers that produce a lot of nectar, which is why they are popular with beneficial pollinators. The berries may also be partially pollinated by the wind.[5]

Site and Soil Conditions: Although blackberries can be found growing in the wild in many settings, they are most productive in full sun. Brambles can survive in average-quality soil as long as they get consistent moisture through the fruit ripening period, but they will thrive and be more productive in well-drained soil enriched with organic compost or decomposed leaf mulch. The canes often develop fungal disease problems and ultimately die in wet, soggy soil conditions.[6]

Managing a Blackberry Patch: Methods for growing blackberries range from allowing the plants to form a wild bramble patch, to formally training the canes on a trellis system. Below are a few methods we find best for dealing with the canes.

Cane Thinning: If a moderate number of canes are thinned out each spring as they emerge, the remaining blackberry canes will ultimately be thicker and produce larger fruit. You can do this by keeping the healthiest and thickest two to five stems for each cane system and cutting all the other weak, thin canes to the ground. Be aware that a crowded blackberry patch will also hide more pests, so regularly cutting away dead canes will make it easier to manage brambles and discourage diseases and pests.

Semi-Organized Wild Blackberry Patch: If you are lucky enough to have a wild patch of sweet blackberries on your property, you can manage them in order to encourage more productivity. By their nature, brambles are wild and messy, which makes it difficult to get to the fruit without being scraped by thorns. Thorny tangles can be partially alleviated by using gardener's twist-tie wire to tie sections of the bramble to one another in order to create handy access to the fruit.

Berry cane management is best implemented in early spring before the foliage has opened up. You can create small openings for better fruit picking every 4 to 6 feet (1.2–1.8 m) apart by pulling berry canes to the side and attaching them to other canes with the ties.

Support System or Trellis: Since the ends of blackberry canes and fruit often wind up in the dirt, a trellis support is the standard method to manage berry brambles. A series of 6- to 8-foot (1.8–2.4 m) high posts are buried 2 feet (61 cm) deep in the soil at a distance of about 6 to 10 feet (1.8–3.1 m) apart. These posts hold horizontal 4- to 6-foot-wide wooden crosspieces that support galvanized wire stretched out between the posts to accommodate the arching canes. As the canes grow each year, fasten them onto the galvanized wires at a convenient height for picking with twist-tie wire so that the berries ripen in full sunlight for better-quality fruit.

Shortening Canes: Blackberry canes can grow from 6 to 18 feet (1.8–5.5 m) in one season, so growers may want to shorten their lengths through pruning. Each spring after new shoots have grown out 2 feet (61 cm) in length, cut the canes back to the first two to three sets of leaf buds. This will result in thick, shorter canes that are upright and one-third to one-half of their original length; these tend to produce bigger fruit.

Hardiness: Blackberry plants have a wide range of hardiness, with many plants rated hardy to zone 5, or −20°F (−29°C) or colder. When selecting berries for your garden, you should plant cold-hardy varieties that are appropriate to your growing region.

Fertilization and Growth: Blackberry canes have shallow roots that do not like to compete with weeds, so regular weeding will allow the roots to spread out and develop into larger fruiting systems. Blackberry roots like to stay cool and moist, so a spring and summer application of compost will not only increase fruit yields but also help prevent their roots from completely drying out.

Cultivars: Listed below are a couple of the contemporary cultivars with thorny canes that produce high-quality fruit. Both have been bred for disease resistance.

'Black Satin': Productive canes that grow huge 2-inch (5.1 cm) berries with good flavor.

'Obsidian': High-quality berries, bred to produce extra-large crops.

Several thornless cultivars that have been developed over the past two decades including both of the following.

'Chester Thornless': Disease-resistant canes bred for large, sweet fruit.

'Triple Crown': Vigorous-growing smooth canes that consistently produce flavorful berries.

Related Species: Blackberries are closely related to red and black raspberries. There are over two hundred and fifty species in the *Rubus* genus.

Propagation: Blackberries can be propagated in many ways, but the simplest method is through tip layering. In spring simply bury the front end of a healthy berry cane firmly into the soil to a depth of about 2 to 4 inches (5.1–10.2 cm), and it will generally form roots after a year.

Pests and Problems: Blackberries have some minor pest problems such as stink bugs, wasps, and birds. All parts of the plant, including the fruit, can develop gray mold (*Botrytis cinerea*), but the worst problem for all blackberries is orange rust disease. There are two types of orange rust fungal pathogens: *Arthuriomyces peckianus* and *Gymnoconia nitens*. While these are rare, they do occur across the United States and can fatally affect wild berries. To avoid this, grow all cultivated blackberry varieties as far away from wild brambles as possible to reduce the risk of spreading pathogens, which are windborne between plants.

Blackcurrant

Ribes americanum, R. nigrum

American Blackcurrant, European Blackcurrant

> *Notwithstanding the lack of popularity of the black currant in America, we might do well to cultivate it more than we do. It undoubtedly possesses more value than we accord to it.*
>
> —FRED W. CARD, 1914[1]

Native to northwest Asia as well as northern Europe, the blackcurrant (*Ribes nigrum*) is popular and valued throughout Europe as a dessert fruit for a wide array of culinary uses. The American blackcurrant (*R. americanum*) is broadly distributed across the northern parts of the United States east of the Rocky Mountains from Nova Scotia to Virginia.

American blackcurrant is a humble fruit that is rarely cultivated as a desirable food crop and is thought of primarily as a forage for wildlife. Although European blackcurrant is larger and slightly sweeter than the American, they both possess that distinctive, musky flavor that is the most characteristic feature of the fruit.

Though not popular in the United States, European blackcurrants (*R. nigrum*) have been popular for hundreds of years for fresh eating, cooking, and medicinal uses.[2] Indeed, there are many countries in Europe whose regional culinary delicacies require blackcurrants as the key flavoring. Crème de cassis is a distinctive liqueur enjoyed throughout northern Europe that is made from the fruit, and there are large-scale European orchards dedicated to producing berries for currant juice and juice concentrates. In several eastern European countries, the leaves are used to make traditional teas and for pickling cucumbers, while Scandinavian countries use the berries to make traditional baked desserts.

There are several black-fruited currants native to the United States, but for the purposes of this chapter we are focusing on the American blackcurrant (*R. americanum*), which has had a humble and legally problematic history. The fruit was originally collected as a wild food by Indigenous tribes and early colonial settlers; by 1925 there were no major commercial blackcurrant orchard plantings and few farmers offering the fruit for sale.[3] But in 1912 the American blackcurrant went from being a neglected wild berry to a legal fugitive when it was discovered that *R. americanum* was a host species for white pine blister rust (*Cronartium ribicola*). This destructive disease uses both the blackcurrant and some of the native pine trees to complete its life cycle, which causes no harm to the blackcurrant plants but can be destructive to certain native pine trees, specifically eastern white pine (*Pinus strobus*). Because of the historical importance of white pine in the US lumber industry (and the lack of a currant fruit lobbyist in Washington), federal and state governments introduced restrictions on the import, planting, and cultivation of *Ribes* species, which banned all currants from being commercially grown in the United States in order to protect white pine trees.[4] While most of the federal laws were rescinded in 1966, some restrictions remain at the state level, and a number of states still prohibit blackcurrant bushes from being sold. So while this may be one of the reasons for the blackcurrant's lack of popularity, the fruit's pungent taste combined with the small, dried, but edible, calyx ("tails") on the ripe berries may also be contributing to its absence in both mainstream and farmers markets.

For over fourteen years we have grown three American blackcurrant bushes near white pine trees that have never been affected by the disease, but caution should be taken if you live in an area with a large population of pine trees. If you want to grow blackcurrant and it is not prohibited in your state, try planting a limited number of disease-resistant varieties from the list in the "Cultivars" section and plant the shrubs far away from any pine trees (if possible).

'Crandall' clove currant flowers.

Attractive flowers of blackcurrant.

Blackcurrant bush.

'Titania' strigs.

Growth Difficulty Rating: 1. The American and European blackcurrant have few, if any, major problems, though the latter can get a few pests and mold issues that may require minor control. But in all other aspects of cultivation, blackcurrant bushes are generally carefree plants in most environments that get consistent moisture year-round.

Taste Profile and Uses: This black-skinned fruit is technically a berry containing three to twelve tiny edible seeds and ranges in size from ¼ inch (6 mm) (*R. americanum*) to ½ inch (1.3 cm) (*R. nigrum*), with a few select cultivars producing extra-large fruit.

Blackcurrants have a unique aroma with a musky flavor that can range from mild to fairly strong. The European cultivars tend to produce berries that are larger with less of the musky characteristic. Blackcurrants can be eaten raw but are generally used as an addition to baked goods and made into a variety of desserts, jams, syrups, alcoholic beverages, and fruit juices. Because of their tart and astringent flavor characteristics, blackcurrants are also used as an addition to savory meals.

Currants are not only a flavorful fruit but are very healthy as a source of antioxidants due to their dark pigmentation, and are full of anthocyanins, vitamins, and minerals; they contain one of the highest amounts of vitamin C of all temperate-climate fruits.[5]

Plant Description: Both species of blackcurrants are deciduous, multi-stemmed shrubs with either upright or spreading branches that vary in growth habit depending on the cultivar. In the wild, European bushes can reach 8 to 10 feet (2.4–3.1 m) in height, but most of the cultivated varietals grow 3 to 5 feet (0.9–1.5 m) tall, with American blackcurrants also reaching about the same height. Currant branches are thornless and leaf out in early spring with 1½- to 2-inch (3.8–5.1 cm) long maple-like leaves with serrated margins, growing from buds that are alternately arranged along the stems. In the fall the leaves of *R. americanum* are quite colorful!

Flowers: American blackcurrant bushes produce drooping racemes (flower clusters) of blossoms that are about 1 to 3 inches (2.5–7.6 cm) long. Each raceme has five to fifteen small, yellow flowers that dangle down like little bells and emerge shortly after the foliage. The European blackcurrants are a little less ornamental, with simple soft, pink-colored flowers. Currant flower clusters (and their fruits) are produced on strigs that make this an attractive addition to any garden.

Pollination Requirements: There is little literature on the pollination requirements of the American blackcurrant. We have three mature shrubs in our gardens that have grown into one another and produce enough fruit for both us and the birds. European blackcurrants are considered self-fertile, but we think it is best to have at least two bushes of either species if you want to have an abundance of fruit.

Site and Soil Conditions: The bushes are best grown with consistent moisture throughout the growing season and planted in partial-shade conditions; though blackcurrants are adaptable plants and can produce fruit in almost complete shade, they will be more productive with more sun. In full sun the shrubs may be scorched during dry periods in the hotter days of the summer, though this may be helped by supplemental watering and mulching the plants' root systems to keep them moist and cool. European blackcurrant requires well-drained soil, but in the wild, American blackcurrant is found in many types of growing conditions, including wetland areas, so the shrubs may be tolerant of wet soil as long as the plant's roots are not constantly standing in water.[6]

Blackcurrants are best suited to northern climates with shorter and cooler summer growing seasons than in the hotter southern parts of the United States, where the shrubs may decline over time from excessive heat or protracted drought.

Hardiness: Blackcurrant bushes are very hardy to growing areas in zone 4, or −30°F (−34°C), but many cultivars are native to the northern parts of Europe such as Russia, whose long cold winters go as low as −40°F (−40°C)! American blackcurrants are also native to cold areas with long snowy winters with a range extending up into Saskatchewan in Canada.

Fertilization and Growth: Blackcurrants are adaptable shrubs that grow in many types of soil conditions, and while it is true that they will survive with little effort, a gardener will be rewarded with consistently better fruit harvests by enriching the soil around the bushes with a thick, yearly application of aged manure mixed with a handful of fireplace ash as a mulch to keep their shallow root systems cool.[7]

Cultivars: There are few if any cultivars of the native American blackcurrant. However, because of the long historical popularity of European blackcurrants as a food and for medicinal preparations, over fifty named cultivated varieties have originated in northern European countries.[8] Although the plants are not widely planted and cultivated in the United States, there are good sources in the mail-order nursery trade, and to a lesser degree local plant nurseries. Below are a few cultivars regarded as being immune or resistant to white pine blister rust.[9]

'Crusader': A high-quality, rust-resistant Canadian variety that produces large fruit and is best planted with another blackcurrant for good pollination.

'Minaj Smyriou': A productive Bulgarian variety that is rust-resistant and produces large-sized fruit.

'Titania': A vigorous-growing and productive Scandinavian cultivar that is rust-resistant with a reputation for producing large berries that have a milder taste.

Related Species: There are between 125 and 150 species in the *Ribes* genus, from which there are 10 species of blackcurrants found throughout the world, growing mainly in the cooler temperate parts of the Northern Hemisphere. Many of the species that produce edible berries are also cultivated for their attractive ornamental flowers.[10] 'Crandall' clove currant (*R. odoratum*) is a late-nineteenth-century American cultivar that is widely planted for its larger fragrant flowers and milder-tasting large fruit. 'Crandall' is also regarded as being more resistant to the white pine blister rust. Red flowering currant (*R. sanguineum*) is a blackcurrant shrub native to the western United States that is cultivated for its beautiful scarlet-pink flowers and edible fruit.

Propagation: Plants in the *Ribes* genus are among the easiest to start from cuttings. To propagate blackcurrants from cuttings, select either one-year-old stems that are firm or two-year-old hardwood stems that are about 6 to 10 inches (15.2–25.4 cm) long and close to the thickness of a pencil, from the late fall through the early winter. Remove most of the leaves, keeping only a few on top, and plant in a pot with most of the stem in soil deep enough so that only a few inches stick out. Provide some winter protection from the cold, and in the following spring the buds at the top of the cuttings should begin to sprout.

Pests and Problems: We have never had mildew problems on our American blackcurrant shrubs, but it is occasionally a problem with European plants. It can be alleviated by thinning the bushes for better air circulation.[11] Branches that are four years or older become unproductive stems with less fruit, so judiciously removing a few of the older branches each season will encourage newer flowering stems. That way bushes stay productive, and you'll have the added benefit of alleviating potential mildew problems.

Boysenberry

Rubus ursinus × idaeus

> *In 1935 the Boysen (Boysenberry), very similar to the Young, was introduced. Its origin is unknown.*
>
> —George Darrow, 1937[1]

Pomological history is filled with many types of novelty fruits that were the result of hard work done by great botanists or fruit breeders on plant hybridization. But it is also filled with lucky discoveries in farm fields, and the outright stealing of another plant breeder's labor, and putting a new name on an old plant as a marketing scheme. The boysenberry is a great example of a wonderful fruit that has a checkered and dubious origin, one that involves so many horticulturalists "borrowing" the plants of other growers that the fruit's origin is at best murky and almost impossible to determine.

Boysenberries are believed to have the combined messy gene pool of the blackberry, raspberry, and loganberry (which is already a hybrid).[2] But the exact species that were used to make boysenberry are unclear, and it may in fact have been a plant taken (legally or otherwise) from the famous plant breeder Luther Burbank and made its way to John Lubben's farm in Napa, California, where it was called 'Lubbenberry.'[3]

Flower bud.

Beautiful blossoms.

Boysenberries ripening.

Berries and foliage.

Rudolph Boysen was farming on Lubben's property in the 1920s and started to hybridize the berries (or just used Lubben's original plants). He transplanted them to a new farm located in Anaheim, California, in 1923, and several of the plants started producing large-sized berries with a good flavor that he claimed to have bred.

Then in the late 1920s, the great berry breeder George Darrow of the USDA and a California berry farmer named Walter Knott tracked down some plants from Rudolph Boysen's failed farm and "discovered" a few surviving plants overgrown with weeds. These berries were introduced to the public in 1932 under their new name, boysenberry, and sold at the popular California roadside farm stand and tourist amusement park called Knott's Berry Farm.[4]

In the middle of the last century, the boysenberry became one of the most popular berries in the United States, but unfortunately by the 1960s the popularity of the fruit began to decline because farmers got more than a few headaches dealing with the crop. The berry canes' trailing habit made the plants difficult and expensive to manage; the plants were susceptible to disease in the coastal areas of California; the fruit had soft skins that leaked juice and only had a short shelf life; and the final nail on the lid of their proverbial coffin came when cheaper blackberry hybrids from home and abroad replaced boysenberries in produce aisles.[5] Thus it was the sad fate of the boysenberry to be replaced by other hybrid blackberries that were more supermarket-friendly. But for most home gardeners and berry lovers, the fact that these berries have a short shelf life and a thin skin that leaks juice is not important if your main goal is to stand in front of a bramble and eat sweet berries with stained fingers, with enough fruit left over for making boysenberry pies.

Growth Difficulty Rating: 2. Boysenberries are easy to grow and only have a few minor problems, but all mature berry cane systems will eventually become unruly, and their growth will need to be managed for easy access to the fruit.

Taste Profile and Uses: The fruit is not a true berry but is classified as an aggregate of drupelets. Each fruit is composed of small drupelets, tightly clustered around a solid core, each with a single seed.[6] Boysenberries are a dark purple-maroon color with occasional red areas and grow to about an inch (2.5 cm) long.

Fresh boysenberries are soft and thin-skinned and prone to leaking juice when fully mature, which makes them delicious, juicy, and slightly messy for fresh eating. They are also highly perishable and need to be used within a day or two of harvesting. Dead-ripe boysenberries taste like a very sweet blackberry with a tangy finish, which is what makes the fruits so delicious. We like to eat them standing in front of the plant, but boysenberries are also great for making jams, jellies, pies, fruit juice, and wine.

The fruit ripens over a period of two months in the summer and can be frozen for later use. Like other dark berries, these contain anthocyanins due to their blue-purple pigmentation, and have natural antioxidants, making the fruit a good addition to a healthy diet.

Plant Description: Boysenberry root systems produce vigorous, sprawling canes that grow 6 to 12 feet (1.8–3.7 m) long and require some sort of support so that, when loaded with fruit, their branches are not lying on the ground. The thornless varieties of boysenberry have smooth canes that produce good crops each growing season but are less vigorous and productive than the standard thorn-covered canes.[7] The 3- to 4-inch (7.6–10.2 cm) leaves on boysenberry varieties have toothed edges that are produced in sets of three with a deep green color on top and a lighter green on the bottom.

Life Cycle of Berry Canes: All bramble stems are biennial. In the first year, a primocane (new cane) grows from a bud in the roots into a long, slender, curving stem with leaves but no flowers. In the second year, the original cane, now called a floricane, develops a second flush of leaves with flowers that ultimately produce the fruit. After fruiting the individual floricanes will die at the end of the growing season. Canes often thrive for eight to ten years, and after this length of time the root systems begin to decline for several years until finally dying off. The longevity of berry brambles can be increased with a yearly addition of well-aged compost or leaf mulch to the soil surrounding the roots.

Flowers: Boysenberry canes produce small flowers that are an inch (2.5 cm) wide with five white petals occasionally tinged with pink.

Pollination Requirements: Boysenberry canes are self-fertile. The blossoms open in spring and produce large amounts of nectar, which attract bees and other insects.[8]

Site and Soil Conditions: Boysenberries require full sun for the best fruit production, with soil that is slightly acidic and well drained. They will grow in average soil, but rich soil that has been amended with aged compost or aged leaf humus will help them grow vigorously. As with all species in the *Rubus* genus (blackberry and raspberry), the root systems of boysenberry plants can often develop fungal disease problems and ultimately die in soil that is constantly wet.[9]

Managing Boysenberry Canes: Most berry brambles that are healthy eventually grow into large, unruly tangles that require some type of trellis or support system for convenient access to the fruit. In the "Blackberry" chapter we have included a section titled "Managing a Blackberry Patch" that lists several different ways to make a support system for berry brambles.

Hardiness: Boysenberry plants are rated as being hardy to zone 6, or about −10°F (−23°C), but some cultivars are rated with a greater cold tolerance that is listed as zone 5 (−20°F/−29°C), so be sure to check the specific cold tolerance for individual plants.[10]

Fertilization and Growth: Boysenberry canes have shallow roots that do not like to compete with weeds, so regular weeding around the bramble canes will allow the plants to spread without interference. The roots like to stay cool and moist; an application of compost or leaf mulch around the plants each spring will not only help prevent the boysenberry roots from drying out but also increase fruit yields.

Unripe fruit.

Cultivars: These plants have become a popular part of the nursery trade in New Zealand, and the majority of cultivars of boysenberry are being used in that country and cannot be purchased in the United States. There are only a few cultivars of boysenberry available through the nursery trade and in mail-order catalogs, with the thornless varieties being the most widely sold. The two cultivars below are generally available.

'Boysenberry': A cultivar sold as the traditional boysenberry, created by Rudolph Boysen in the 1920s, whose thorn-covered canes produce large, sweet-tasting fruit that is a hybrid of several different species.

'Thornless Boysenberry': A thornless cultivar with vigorous growth that produces good crops of large, flavorful berries and that are fairly pest-free.

Related Species: The large *Rubus* genus includes hundreds of species on almost every continent such as the domesticated European raspberry (*R. idaeus*).

Propagation: As with most plants in this genus, the easiest way to propagate the boysenberry is to bury the tips of the canes in a few inches of soil, which will establish a root system when left in place for one or two seasons. Then you can cut the end of the cane around 6 to 8 inches (15.2–20.3 cm) above the base of the plant; making sure to include a few sets of leaves. These can be potted up with soil for another year, which will allow the young plants to form larger root systems, for planting in a new location.

Pests and Problems: Boysenberry cultivars are generally pest-free, and our plants have never been bothered by any major pests, although wasps do like the fruit and occasionally drill small holes into the berries. Birds and other wildlife sometimes take the ripe (and unripe) fruit off the canes. All parts of the plant, including the fruit, can develop gray mold (*Botrytis cinerea*), but this is usually only a problem during periods of heavy rain around the time that the fruit is ripening. The worst disease that all plants of the *Rubus* genus can get is orange rust (*Arthuriomyces peckianus*), which is always fatal for berry brambles. This is a rare disease, but it can occur in many parts of the United States. To avoid this you should site all boysenberry plantings as far away as possible from any wild blackberry or raspberry plants that are native to your growing area to reduce the risk of spreading the pathogens between plants.

Che

Maclura tricuspidata

Che 'Seedless Female'

> *... the plant itself, although interesting both economically and botanically, has no special merit for the garden. Its leaves are much used in China for feeding the silkworm on, being considered as good as mulberry—to which also it is closely related.*
>
> —W. J. Bean, 1922[1]

Though che is a little-known fruit, it has undergone many name changes for both its formal Latin reference and its common name. Che's official Latin name was originally *Cudrania triloba.* Then it was reclassified as *C. tricuspidata* and appeared as such in pomological literature until recently, when it was moved to the genus of the Osage orange (*Maclura pomifera*).[2] Now it is called *M. tricuspidata*, yet that simple change does not even compare with the dizzying array of common names for che. Although this is not a complete list, here are some of the names for the fruit: Chinese che, Chinese mulberry, cudrang, red Chinese mulberry, wild mulberry, Mandarin melon berry, and silkworm thorn.[3]

Many of these names might lead you to suppose that the flavor of che has similar qualities to mulberry (a close relative), which it does. But the best thing about che is that its taste is also like another of its relatives, the fig. When you combine these two qualities, and mix in the flavor of watermelon and lychee, then you have some notion of what the fruit tastes like.

Che seems to have originated in several parts of eastern Asia, and has naturalized in Japan. It was the historic silk trade that brought mulberries, as well as the che fruit, into global commerce.[4] Che was introduced into England and other parts of Europe around 1872, and then into the United States in 1909, by the famed British plant hunter E. H. Wilson.[5] In China and other parts of Asia, the fruit is sometimes found in markets, but it's not commercially cultivated in other parts of the world.[6] Since the 1980s che has started to appear in plant catalogs and pomological literature, which we think is a good thing because the trees are beautiful. In autumn our che tree is truly striking, and is often mistaken for a *Cornus kousa* tree since dogwood trees also have showy red fruit that is similar in shape and size. Che fruit is very

Ripe fruit.

decorative, especially when the fruit changes from pale cream to vivid red, and visitors to our gardens always ask about this tree.

Growth Difficulty Rating: 1. Che trees are easy to grow with almost no pest problems.

Taste Profile and Uses: Each che fruit is an aggregation of multiple small drupelets clustered together to form a single fruit around a stem. Our seedless female che fruit is approximately ¾ to 1 inch (2–2.5 cm) in diameter, round with a bumpy surface and an appearance that is somewhat like a lychee. The fruits turn a pink-red color as they ripen, and are an attractive deep maroon-red color with small black marks on the outer skin surface when fully ripe in the fall. Cold night temperatures may cause some of the fruit to fall from the trees.

The nice thing about che is that the fruit ripens unevenly over a period of four weeks and persists on the tree for a long time when it is fully ripe. After waiting for six years, our grafted self-fertile 'Seedless Female' che tree finally got fruit that fully ripened and didn't prematurely fall off the branch. The inner flesh of the fruit is also colored a deep red and has a soft texture. The flavor was much sweeter than we thought it would be, and well worth the wait. An unripe che is hard and flavorless, but when dead ripe, the fruit is soft and similar to mulberry without the sharp acidity of some mulberry cultivars. Che also has an intense sweet flavor like watermelon and lychee, with some notes of ripe fig.

Plant Description: A deciduous, multibranched tree that eventually can grow to about 25 feet (7.6 m) in height, but often remains a broad, spreading shorter tree. In the first years of growth, che trees can sport branches with sharp thorns, but these usually disappear as the plants mature in age. The alternate leaves are a rich deep green color with variable shapes, sometimes entire (smooth edges with no indentations), or with two or three lobes that end in pointed or rounded tips. They have a waxy covering called a *cuticle*, which gives them a leathery appearance.

In the wild, female trees are often larger than male, but there seems to be a lot of variability in all aspects of size and shape. Che often grows into an unruly form with multiple suckering stems, which is why nurseries graft che onto Osage orange rootstock to create a neat upright tree.

Flowers: The flowers are yellowish green and considered inconspicuous since they are the size of a small pea. They are borne on new

Che tree.

Flower buds.

Thread-like female flowers.

Ripening fruit.

wood on short peduncles (stems) and look like teeny versions of the fruit. On our grafted 'Female Seedless' specimen, the flowers develop many small stigmas, giving them a hairy appearance.

Pollination Requirements: Wild seedlings of che are typically dioecious, with trees having all-male or all-female flowers that are wind-pollinated. To get fruit usually requires separate male and female plants. These will produce larger-sized fruit with seeds inside.

Like all aspects of che, flowers also seem to demonstrate variability. There are references in pomological literature about che flowers that suggests that some trees are one sex but bear a small number of flowers of the other sex, thus being somewhat self-fertile. There are also diverse opinions about the possibility of female plants being able to produce fruit parthenocarpically (able to set seedless fruit without a pollinator).[7] Che belongs to the Moraceae family, which contains other fruiting plants that do exhibit parthenocarpic fruiting, such as mulberry and fig, so this may also be happening with some che trees. Given the ambiguity involved with che pollination requirements, our advice is to purchase self-fertile grafted trees. We have heard that the fruit resulting from the pollination between a male and a female tree is full of seeds and unpleasant to eat.[8] In our arboretum there is a single self-fertile female tree that produces an abundance of delicious seedless fruit in October when most fruiting plants are done.

Site and Soil Conditions: This plant requires a growing site that is in full sun. *Che* supposedly means "stony ground," which may be a reference to the tree's natural habitat of poor, dry soil.[9] But che seems adaptable to diverse soil conditions as long as it is not planted in overly wet soil. At Hortus the tree has thrived in poor-quality clay soil that was amended with compost in a full-sun planting site. There are numerous references to che being drought-tolerant and surviving in the drier western parts of the United States, as long as the trees get enough water to get them established. After this the trees produce fruit and survive in dry conditions as long as they get some water every few weeks throughout the growing season. Our gardens generally get consistent rain every week (or so) with dry periods in the middle of summer, but our tree has also produced fruit in years that experienced an unusually large amount of rain.

Hardiness: Che has a wide range of adaptability and is rated to survive temperatures down to −20°F (−29°C) or zone 5, as well as thriving in subtropical growing conditions. Although these plants may survive in zone 5, the shorter growing season, or an early frost, may limit the fruit's ability to ripen on the tree. Which is why we think this is a better choice for a zone 6 environment (–10°F/–23°C).

Fertilization and Growth: Che is able to grow in poor soil but will really thrive in rich, fertile, well-drained soil. Che trees like a hot, sunny location. The unripe fruit will fall off the tree in periods of extended drought, so we feel that younger trees may benefit from supplemental water during the dry periods of summer to help the fruit ripen.

Cultivars: There are relatively few cultivars of che available in commerce, and not a lot of research on which ones are the best for producing fruit. All of the cultivars listed below are self-fertile female trees.

'Che Seedless': A vigorous seedless tree cultivar discovered in Ivy, Virginia, that produces sweet, high-quality fruit about 1 inch (2.5 cm) wide.

'Darrow': One of the oldest che trees introduced into the United States, originally collected by the USDA plant breeder Dr. George M. Darrow.

'Norris Seedless Female': Discovered in the 1930s by the nurseryman Cliff England growing in a fruit orchard in the Tennessee Valley Authority's Norris Dam area, with ½-inch (1.3 cm) sweet fruit.

Related Species: Che is in the family of Moraceae, which includes fruiting plants that are scattered all over the world. Its relatives include the North American native tree Osage orange (*M. pomifera*) and mulberry (*Morus* spp.), as well as tropical relatives like breadfruit (*Artocarpus altilis*) and jackfruit (*A. heterophyllus*).

Propagation: Che is traditionally grafted onto Osage orange rootstock because of its close relationship to that species. For ease of cultivation, this rootstock is a good choice because it eliminates root suckering and produces an upright tree form. Che can also be grown from seeds, but the trees will take many years to mature and will be unsexed plants.

Pests and Problems: Che does not seem to be bothered by any major pests. We have had Japanese beetles chew on the leaves, but they do not do significant damage.

Chinese Kiwi

Actinidia arguta

Bowerberry, Hardy Kiwi, Kiwiberry, Tara Berry

> A. arguta *is excellent for covering arbors, trellises, walls and the like and is remarkably free from insects and fungi.*
>
> —L. H. BAILEY, 1930[1]

Hardy kiwi (*Actinidia arguta*) may be familiar to the reader since the small fruits are sometimes sold in grocery stores and marketed as kiwiberries. There is much confusion about this fruit's common name; it is often mistaken with its sister species *A. kolomikta* (see the "Arctic Kiwi" chapter), so for our purposes we will refer to it as Chinese kiwi. This kiwi is also a rambling vine, even more vigorous than arctic kiwi, and it can reach 100 feet (30.5 m) tall by growing into the forest canopy of its native habitats in eastern Siberia, Manchuria, northern China, Korea, and Japan.[2] These fast-growing vines are similar in many respects to other kiwi species, with individual plants being either male or female. Chinese kiwi foliage has beautiful heart-shaped leaves, but it lacks the magnificent variegation of the arctic kiwi species.

Chinese kiwi was grown and distributed by several nurseries after it was introduced into this country near the turn of the century, and many of the original plants are still growing at public and private gardens.[3] At the Arboretum we planted a male and female plant, the same year we planted the arctic kiwi vines, and it became clear that while the latter were precocious, setting flowers and fruit within three years, the Chinese kiwi vine produced lots of green leaves and nothing else. After a decade of growing it, with still no flowers forming, it became a hotly contested plant in the garden with threats to pull it out and to plant something more productive. That threat must have worked because the next year in late spring there were a handful of small white flowers. The following year there were even more flowers, but still no fruit. It was thirteen years after planting the vines that we got to taste Chinese kiwi, and it ended up being well worth the wait: The large delicious fruits have an even more complex taste than the arctic kiwis, with notes of pineapple (and sometimes mint). For most situations we still advise folks to grow arctic kiwis, but if you have a large area (and a lot of patience), we think growing Chinese kiwi is a good choice, with fruits ripening about six to eight weeks later than arctic kiwi.

Growth Difficulty Rating: 2. Kiwis are easy to grow with no major pest problems, but because the vines can grow up to 30 feet (9.1 m) a season, they will require a strong support system, such as a trellis, to keep the fruit manageable. Given their long lateral growth, the vines will eventually require some pruning to stay within the boundaries of most garden settings.

Taste Profile and Uses: Chinese kiwis have a similar taste profile to the supermarket variety as well as the arctic kiwi, with the addition of having a more rounded, full-bodied sweetness with notes of acidity that remind us of pineapple and mint. We actually preferred their more complex flavor to the arctic kiwi! Like the arctic types these are grape-sized fruits that grow 1 to 1¼ inches (2.5–3.2 cm) without the brown fuzz, so that the whole fruit can be popped into your mouth, no peeling required. The fruit is technically a berry, and if you cut it in half you will see a ring of teeny seeds just like the store-bought types. The fruits are more rounded than the arctic species and hang from red petioles (leafstalks) in small clusters that range in color from dull green to brownish green, with some cultivars having a purple-reddish cast that is highly ornamental. The fruits do not fall off when fully ripe like arctic kiwi, so they can persist a long time on the vines, well into

Male flowers.

Female flowers with fruits forming.

Ripening fruits.

The ornamental peeling bark.

the end of October in the gardens. The fruit is best enjoyed right out of hand but are also great in smoothies, desserts, and salads. Chinese kiwis store particularly well refrigerated, but we find freezing them is best for long-term usage.

Plant Description: Chinese kiwi is a large, sprawling, perennial twining vine that can grow up to 30 feet (9.1 m) a season. Both the male and female plants have simple, shiny, alternate dark green heart-shaped leaves that grow 3 to 6 inches (7.6–15.2 cm) long and are much more finely toothed than arctic kiwi, without any color patterning, which is a good way to identify the species. Their bright red petioles have a lovely quality and are also a key way of confirming the plant as Chinese kiwi. As the vines mature, the plant's trunks and mature branches exhibit exfoliating bark,

Leaves and flower buds.

which makes them quite showy in winter. The vines will definitely require a strong support structure such as a trellis, wall, or fence that can bear the weight of a large, established fruiting vine.

Flowers: Chinese kiwi blooms after arctic kiwi with flowers that are slightly larger and have waxy, whitish green petals. They are often obscured by the lush foliage, but their heavenly fragrance perfumes the air. Each female flower has a tiny green ovary (while the male does not), topped by up to twenty stigmas (pollen receptacles). This is the surest way to distinguish between male and female plants, which are occasionally mixed up by nurseries and incorrectly labeled.

Pollination Requirements: Since this vine is dioecious, with plants primarily having all-male or all-female flowers, you will need one of each to get the delectable fruit. Some plants can produce both male and female flowers, and others have been reported with perfect flowers (having both male and female structures). Chinese kiwi may even change sexual expression from year to year, which may be why the vines are not always consistent in their fruit production.[4] It's important to source plants that have been properly identified. One male plant can pollinate up to six female vines. Many species of bees seem to love this plant, doing the work of pollinating the flowers.

Site and Soil Conditions: We have found that Chinese kiwi vine is an adaptable plant that prefers nutrient-rich soil that is well drained, but it will grow in average soil conditions if provided with some compost every year. At the Arboretum our vine seems to tolerate infertile soil and was not given additional compost or mulch for many years, which may have contributed to the vines taking so long to produce their first flowers. Our plants are sited in a part-shaded area that gets afternoon sun, which may also have caused the long wait for the fruit, because these vines are more productive in full sun.

Hardiness: Chinese kiwi vines are hardy to zone 4, or −30°F (−34°C). If you are gardening in the coldest parts of the United States, it is advisable to protect young kiwi plants, which have shallow root systems for a few seasons of growth, with a thick protective layer of mulch, but after that the vines are usually resilient enough to survive harsh winter conditions. Flowers can also be killed by a late-spring frost.

Fertilization and Growth: Kiwis are shallow-rooted vines from forest and scrubland areas, and like all plants that are native to woodland environments they would benefit from soil enrichment with a yearly application of leaf mulch. But if you don't have leaf mulch, garden compost or well-aged manure can also work. We also recommend a spring addition of compost after heavy fruiting years. We have seen this plant growing in a full-sun aspect and it seems almost uncontrollable, with vines being weighed down by hundreds of pounds of these small fruits.

Cultivars: Lots of varieties of this plant are offered in mail-order catalogs. Here are a few of the more popular cultivars.

'Ananasnaya' ('Ana' or 'Anna'): A Russian variety whose name means "pineapple-like," referring to its flavor.

'Dumbarton Oaks': A cultivar with large-sized fruit discovered at the Dumbarton Oaks historical site in Washington, DC.

'Hardy Red': A heavy bearer with red skin and red flesh.

'Issai': The only cultivar that is considered self-fertile, though the fruits are on the smaller size.

Related Species: Pilosula kiwi (*A. pilosula*) is a striking Chinese relative with small pink flowers and attractive narrow variegated foliage.

Propagation: Chinese kiwi is commonly propagated with firm softwood cuttings about 1/4 inch (6 mm) thick that are taken in early summer, or by using hardwood cuttings potted up in soil in the early part of winter and grown out in a shady location for a couple of years. The easiest way to produce new plants is by layering stems in the soil for two growing seasons until they have produced enough of a root system that they can be removed from the mother vine and replanted in another garden spot.

Pests and Problems: We don't see too many pests bothering this vine. Squirrels, chipmunks, and birds like the fruit and will take the berries. Recently, Chinese kiwi is gaining attention as an invasive plant in the United States; its tenacious growth and resilience to low temperatures have allowed it to take over wooded locations and may have a negative impact on certain forest areas.[5] However, many sources challenge that this is an invasive vine, and they offer up useful insight as to how land-management practices can greatly affect the ability of a species to become established.[6]

Chokeberry

Aronia arbutifolia, A. melanocarpa, A. × prunifolia

Aronia, Black Chokeberry, Purple Chokeberry, Red Chokeberry

> *A small or rarely large shrub; flowers March–June; somewhat subject to blight and borer attack; has been observed to grow on cinder dumps where fruit crops are often heavy.*
>
> —WILLIAM RICHARD VAN DERSAL, 1938[1]

As indicated by the quote above—from a USDA publication on native plants for erosion control and wildlife value—aronia bushes are tough plants that are useful for roadside erosion control and may even survive the potential hazards of an occasional automobile running over them. The different species of chokeberries (*Aronia* spp.) have been cultivated by nurseries as carefree ornamental landscape plants that produce beautiful fall colors and provide edible berries to native wildlife.

For a number of years this genus has been marked by taxonomic confusion, which has resulted in many plants being labeled incorrectly or having multiple names when listed in native plant guides and in the nursery trade. Black chokeberry (*A. melanocarpa*) and red chokeberry (*A. arbutifolia*) have very similar geographic ranges that often overlap. But this is only the start of the confusion for the fruiting shrubs, because a plant called purple chokeberry (*A. prunifolia*) is considered by many to be a hybrid between the two species, while others consider it a variant of black chokeberry.[2] But the confusion does not stop there, for the chokeberry also gets mixed up with plants that are from a completely different genus, like chokecherry (*Prunus virginiana*), which is a wild cherry tree species from the Northeast.

All *Aronia* species are deciduous shrubs with a thicket-forming habit that are native to North America and, depending on the species, have a geographic range from Nova Scotia to Michigan then south to Texas and northern Florida. Although aronia is not very popular as a food in the United States, this humble plant is far more celebrated as a nutritious edible fruit in eastern Europe.[3] In Poland the fruit is immensely popular; that country is the largest grower of aronia as a food crop and exports 90 percent of its total harvest to the rest of Europe.[4]

Aronia is also an overlooked decorative shrub to which the nursery trade has given short shrift due to its common name, but this may (sometimes) be justified if you taste the fruits fresh—it can be immediately apparent why they are called chokeberries. The *choke* sensation is the result of aronia fruit being rather astringent, which is always

'Low Scape Hedger.'

strongest in fresh fruit. Despite our lack of enthusiasm for the fresh berries, we dried some of the fruit in a dehydrator, which improved the flavor and reduced the unpleasant astringent qualities. So although we believe that chokeberry is not one of the best fresh fruits, it still has dozens of uses both dried and cooked, and is definitely worth planting.

Growth Difficulty Rating: 1. Aronia is an easy plant to grow that is adaptable to a wide range of growing conditions.

Taste Profile and Uses: The fruit is not a berry but actually a pome ranging in size from ¼ to ½ inch (0.6–1.3 cm) and has the form of a tiny apple.[5] The color depends on the species: The red chokeberry's color ranges from dull red to a brighter red; black chokeberries start out as a purple-wine color that matures to shiny black; and purple chokeberry develops to dark purple. The ripening times depends on the species, with *A. melanocarpa* maturing from July through August; *A. arbutifolia* and *A. prunifolia* produce ripe berries from September to October, with the fruit often persisting on the bushes into the winter.

Black chokeberry produces the largest fruit and has the highest levels of cancer-fighting antioxidants (anthocyanins and flavonoids) of any temperate fruit; it is listed as containing five times more antioxidants than cranberry and blueberry.[6] Because chokeberry has an astringent flavor, the fruit is generally not eaten out of hand but is often used to enhance the flavor and health properties of other fruit juices. Aronia is used in combination with other sweeter fruits to add a tart flavor to jams, jellies, syrups, teas, and wines, as well as adding a natural rich purple-red color. Freezing the fruit is also supposed to reduce astringency, but we prefer to use the berries dried, which enhances their sweetness and makes them a good addition to granola and trail mixes. When fully ripe, chokeberries have the same sugar content as a grape or sweet cherry, but this is offset by its astringent properties.[7] Since there are more antioxidants in aronia than any other northern temperate fruit, they are now considered a superfood and are starting to get the attention of farmers and fruit growers.[8]

Plant Description: This member of the rose family is a deciduous, multi-stemmed shrub that has a habit of growing wider than taller through suckering root systems. Red and purple chokeberry are the tallest shrubs, growing 6 to 12 feet (1.8–3.7 m) in height. Black chokeberry is a smaller bush that forms into a rounded shrub reaching 4 to 8 feet (1.2–2.4 m) tall. Due to hybridization the cultivated varieties grow into a wide range of sizes but generally tend to be less than 8 feet tall. The bushes produce leaves with shiny surfaces that are oval to egg shape ranging in size from 1 to 3 inches (2.5–7.6 cm). Red chokeberry's leaves, stems, and flowers are all pubescent (slightly hairy), while black chokeberry's leaves, stems, and flowers are glabrous (smooth). The purple chokeberry is somewhere between the two, exhibiting a range of different levels of hairiness. As a general rule, all aronia produce beautiful fall color, with leaves that turn scarlet red on both red and purple chokeberry and reddish purple on the black chokeberry. Many cultivars have been bred and introduced for their ornamental autumnal display of purple, red, and orange leaves.

Flowers: In early spring chokeberry bushes produce attractive displays of small white flowers with five petals that are ⅓ to ½ inch (0.9–1.3 cm) wide with pink anthers in the center of each blossom. Although the individual flowers are tiny, mature shrubs will produce blooms in such abundance that the branches are completely covered by the flowers and look very decorative.

Pollination Requirements: *Aronia* plants are self-fertile and pollinated by insects, but this genus of plants has inherited the ability to set fruit without pollination (parthenocarpy). If this occurs the seeds of the fruits will not be viable.[9]

Site and Soil Conditions: Aronia is a very adaptable shrub that isn't very fussy about its soil conditions. It will grow in areas with no nutrients, wet areas such as swamps, and on the edges of forests; we have even seen chokeberry growing out of the soil pockets on the rocky ledge of a mountain cliff. After being watered for several seasons, an established plant can survive drier areas. Aronia plants have been used for erosion control since their roots are thick as well as fibrous and can tolerate drought along with heavy wind exposure.[10] Chokeberries can be grown in a part-shade area, but a full-sun site will yield the best fruit harvests, and all the different species of *Aronia* have a preference for slightly acidic soil that is well drained.

Hardiness: Black chokeberry is hardy to zone 3 or −40°F (−40°C), while red and purple chokeberry are rated to zone 4, or −30°F (−34°C).

Fertilization and Growth: Though this plant can survive infertile soils, the bushes will produce more fruit and vigorous growth if compost or leaf humus is introduced as a mulch around the plant's root system in spring. If some pine needles or pine duff (plant litter) from underneath evergreen plants is available, this will help

Chokeberry flowers.

Developing fruit.

The ripening fruit.

Chokeberry bush.

to create acidic conditions for the soil if used like a mulch. When aronia plants are happy, they will slowly colonize an area by root suckers, so make sure to site these shrubs with this in mind to avoid crowding out other plants.

Cultivars: Black chokeberries are the most common plant in the nursery trade because they produce larger fruit with the best flavor, but many garden centers and mail-order plant nurseries do not differentiate among the three different species and sometimes only offer named hybrids or cultivars that are sold under the name *black chokeberries*. Below are a few varieties that are available and popular for their high-quality fruit.

'Morton': A selection developed by the Morton Arboretum with a compact shape that is floriferous and was marketed in the Midwest under the trademarked name 'Iroquois Beauty'.

'Nero': A short-growing plant that reaches a height of 3 to 4 feet (0.9–1.2 m) with dark blue berries, that was developed in Poland.

'Viking': A vigorous and productive variety that can grow to a height of 6 feet (1.8 m) and was developed in Finland in 1980.

Related Species: Aronia is closely related to juneberry (*Amelanchier* spp.), which is another native plant with edible berries (it's described later in this book), and to *Photinia* spp., which are decorative shrubs native to Asia.

Propagation: Aronia can be started by both seeds and cuttings, but because the root systems of individual plants naturally produce new suckering branches, the simplest way to propagate new plants is by digging up the young branches. You can divide these from the base of the mother plants in early spring and pot them up for a few growing seasons until the root systems get large enough to plant out.

Pests and Problems: Aronia has no major pest issues and tends to get only minor leaf problems that do not affect the overall health of the plant. Birds do eat the fruit, though they aren't inclined to do so until winter, which gives you plenty of time to harvest it.

Root Suckering: All three species have root systems that sucker and will spread out to form a thicket over time. To keep an aronia in the shape of a contained upright bush, the new suckers that form around the base of the plant will need to be pruned away or mowed down.

Cornelian Cherry

Cornus mas

> *A hundred years ago when the number of handsome plants available for American gardens was not as large as it is today the Cornelian Cherry was more often planted here than it is now, and it is doubtful if it can now be found in many American nurseries. Few exotic shrubs, nevertheless, are better worth the attention of northern nurserymen.*
>
> —CHARLES S. SARGENT, 1921[1]

Cornelian cherry has been valued as a food plant in Europe and western Asia since classical Greek and Roman times.[2] It was planted in older European gardens and is mentioned as a medicinal fruit in European herbalist writings from the 1500s.[3] As Charles Sargent indicates in the quote above, Cornelian cherry was introduced to American gardens in the nineteenth century as a spring-flowering ornamental tree. Indeed, the profuse displays of *Cornus mas* flowers could be mistaken for forsythia, an ubiquitous northeastern shrub, because both plants produce abundant sprays of rich golden flowers in early spring when little else is happening in most gardens.

Cornelian cherry is not related to the common cherry, but is an underutilized member of the dogwood family noteworthy for tough, cold-resistant flowers that are followed by attractive, edible fruit. Unfortunately Cornelian cherry fruit often goes unnoticed in public parks, where large old trees regularly produce an abundance of fruit that is eagerly sought by birds and other wildlife.

Cornus wood is very hard and durable, which led to its being valued for centuries for the handles of knifes and meat cleavers. A few years ago on a trip to Ukraine, we saw babushkas on the street corners selling freshly picked seaberry and Cornelian cherry fruits in plastic cups.

Little known in American backyards, this plant can be found in botanical gardens and parks across the country, and we think these well-behaved, slow-growing, shrub-like trees could provide many years of magnificent flower displays and delicious fruit on trouble-free plants.

Growth Difficulty Rating: 1. This plant is easy to grow with no major pests or problems.

Taste Profile and Uses: Cornel fruits are drupes (a fleshy fruit that usually contains one seed), which ripen into deep scarlet-colored berries ⅝ to 1½ inches (1.6–3.8 cm) in size, depending on the cultivar. The fruit can be oval but is typically a beautiful oblong

Early spring flowers.

teardrop shape that is wider and rounder on the bottom, which early gardeners thought looked like carnelian, a scarlet-colored semiprecious stone.

Cornelian cherry must be dead ripe to eat or the flavor will be sour. The taste of the wild fruit can be highly variable, with some trees producing sweet crops while others produce astringent cherries that are better used for cooking, which is why we recommend planting grafted cultivars selected for their high-quality fruit. A good cultivar can produce wonderful fruit that is a delicious blend of tart and sweet, like a rich, flavorful sour cherry. The fruit ripens over a period of weeks from late August to early September, with the sweetest berries dropping into your hand when you tickle the fruit, or they end up falling to the ground near the base of the tree. You can also place a sheet under the plant and gently shake the branches, since ripe fruit is easily released from each pedicle (fruit stem). Besides eating the cherries fresh, you can also use them to make jelly, sauces, juices, or liqueurs, or blend them into cider batches.

Plant Description: These long-lived trees grow slowly and develop into multibranched large shrubs or small trees that reaches 15 to 25 feet (4.6–7.6 m) high, although grafted cultivars will generally grow to about 12 feet (3.7 m). The oppositely arranged oval- or elliptical-shaped glossy leaves are 2 to 4 inches (5.1–10.2 cm) long with the distinct deep-furrowed veins typical of the dogwood family and do not acquire any significant fall color. The bark of old trees is often scaly and exfoliating with gray and brown patterns.

Flowers: Cornelian trees bloom before the leaves open up, which creates a welcoming sight in early spring. The flowers remain intact for two to three weeks and can often tolerate late frosts and still yield a crop of fruit. Although small and delicate, the flowers are produced in such abundance that they often cover entire trees with beautiful sprays of fine golden-yellow blossoms. Young seed-grown plants bloom with only male flowers for many years until they are mature enough to also produce female flowers. We suggest choosing grafted cultivars to expedite fruit production.

Pollination Requirements: Cornelian cherries require at least two trees for good flower pollination, but we have observed old trees planted alone that yield small amounts of fruit.

Site and Soil Conditions: Cornelian cherry is a fairly adaptable plant and can thrive in both sun and part shade but will produce larger crops of fruit in full sun. The trees seem to tolerate a wide range of

The ripe fruit of 'Sunrise' Cornelian cherry.

Incredible early-spring flowers.

Cornelian cherry flowers.

Ripe fruits of the cultivar 'Yellow'.

Fresh-picked fruit.

growing conditions, from rich loam to poor-quality soil, as long as the roots are not constantly waterlogged. These trees have also performed well in the hot summers of the southern parts of the United States. [4]

Hardiness: Cornelian cherry is hardy to zone 4, or −30°F (−34°C).

Fertilization and Growth: Although tolerant of poor soil conditions, Cornelian trees benefit from an occasional mulch of compost to boost fruit production and make the trees more vigorous. In our gardens the three trees grow in heavy clay soil, in part shade, and have received only a moderate amount of fertilization, and still produce a lot of fruit.

Cultivars: There are about a dozen cultivars of Cornelian cherry, a few of which are available in the mail-order nursery trade. Below are a few varieties noteworthy for producing flavorful fruit.

'Pioneer': A large-fruited cultivar introduced by the Central Botanical Garden in Kiev, Ukraine.

'Sunrise': A high-quality cultivar that produces attractive fruit that is sweet and delicious.

'Yellow': A golden yellow–fruited cultivar that is highly ornamental.

Related Species: Korean Cornelian cherry (*C. officinalis*) is closely related to *C. mas* and is a beautiful, small tree with smaller-sized fruit.[5]

Propagation: We have propagated Cornelian cherry by planting hardwood cuttings in potting soil in the late fall. New plants can also be generated by air-layering ¼-inch (6 mm) thick branches with soil packed in plastic bags around branches and held in place with strings.

Pests and Problems: After growing Cornelian cherry for over sixteen years, we have never had a major problem with our trees. However, dogwood anthracnose (*Discula destructiva*) can occasionally strike plants in seasons with large amounts of rain. Birds and chipmunks like the taste of Cornelian cherries, so you will have to harvest fruit several times over a period of weeks or the critters will do it for you.

Cranberry

Vaccinium macrocarpon

American Cranberry, Large Cranberry

> *. . . A statute was passed in 1789 by the governing body of New Jersey forbidding settlers to pick cranberries before 10 October. Those who were caught had to pay a fine of ten shillings, a hefty penalty for people trying to get in before their neighbors by picking berries before they were ripe.*
>
> —JENNIFER TREHANE, 2004[1]

It's pretty amazing that this small fruit, which is too sour to eat out of hand, was seen as a profitable pomological fruit considering that it not only lives but thrives in some of the harshest circumstances, including cold temperatures, boggy and wet conditions, and nutrient-deficient soil.

The use of wild cranberries for food and medicine goes back hundreds of years into the origins of the First Nations tribal cultures. Countless Indigenous tribes gathered harvests of cranberries, which were called *sassamenesh* by the Algonquin peoples and *ibimi* by the

Wampanoag tribes.[2] English colonists to the United States quickly learned from Native tribes that cranberry was a useful wild fruit that could be paired with fresh game in their meals, and centuries later this is still the traditional accompaniment to the Thanksgiving turkey dinner. The popularity of the fruit spread among the settlers of New England, and the first commercial cranberry bog was started by Captain Henry Hall in Barnstable, Massachusetts, in 1812.[3] The first book solely devoted to cranberries was published in 1856, titled *A Complete Manual for the Cultivation of the Cranberry*, by B. Easterwood, to cater to American farmers' interest in a growing industry. By the time of the 1899 US census, no fewer than twenty-seven different states were growing cranberries as an agricultural crop.[4]

Today cranberries are not only a major food crop but have been rediscovered as a superfood by a modern world that has finally caught up to the Native tribal peoples' understanding that the berries can be used for medicinal remedies because they contain many healthful compounds that help reduce urinary tract infections.[5]

Growth Difficulty Rating: 2. Given the right conditions of consistently moist and acidic soil, this is an easy ground cover to grow, one that is fairly pest-resistant and will flourish in the correct soil conditions, that are consistently moist and acidic.

Taste Profile and Uses: Cranberries are ¼ to ½ inch (0.6–1.3 cm) in size and start out white, then ripen to a deep burgundy color in the fall. Cooler autumnal temperatures produce the best deep-red-colored berries, which can occasionally be damaged by the onset of abnormally warm autumn temperatures.[6] All of the berries generally ripen at one time and have a distinctive astringent flavor that is too strong for fresh eating but is one of the great cooking fruits. When cooked with other fruits, they impart a wonderful tangy flavor to sauces and are great for relishes and desserts. The juice itself is sour, but sweetening it allows cranberries to be enjoyed throughout the year, although any sugar added to the juice negates its health benefits. In recent years cranberries that are dried and sweetened have become a popular snack and an addition to granola.

Plant Description: An evergreen ground cover with creeping vines that forms a dense spreading mat by producing woody stems that sprout delicate, ¼-inch (6 mm) long, glossy alternate leaves ranging from narrow ellipses to oval shapes. The woody stems are known as runners or stolons, and can spread many feet from the original plant. In spring the buds in the axils of the leaves produce several vertical shoots called uprights. These upright stems produce the flower buds and slowly grow out and bend down into the tangled mass of the vine over time.

Flowers: Dainty flowers are formed on the upright stems, each producing between five to ten blooms. The buds first appear as little hooks along a pinkish-red petiole; then the petals and the sepals peel back to reveal the stamens, which gives the flowers the appearance of the head and beak of a crane. (The English colonial settlers called them craneberries because the flower was thought to resemble the head of a sandhill crane.[7]) Individually the tiny white to whitish pink blossoms are not overwhelming, but to see a large patch of these unusual flowers is quite beautiful.

Pollination Requirements: Cranberry is said to be self-fertile, but because the plants sold through nurseries are generally small, it is best to plant several individual vines to produce larger harvests.

Site and Soil Conditions: Cranberries are grown commercially in bogs that replicate the plants' moisture requirements and also serve as an easy way to facilitate large-scale fruit production. For the home gardener it is best to try to mimic the fruit's wild growing conditions on a small scale, which is easier than it sounds (see the "Constructing a Small Cranberry Bog" section below). Cranberries can take some shade, but the best fruit production is accomplished in full sun with the vine's root systems growing in acidic soil with a good amount of organic matter and sand. They also can be planted in average garden soil conditions with some acidic amendments, but the plants will need to be consistently watered through the drier parts of the summer.

The best growing conditions for cranberry are when the root systems are constantly moist but the sections of the vine growing out of the soil are not standing in water during the warm months of the year, when the vines are actively growing, because this can cause diseases and can damage plants.[8] Cranberries require some air respiration around their root systems to be healthy during the period in which they are growing. This is why cranberries are only flooded (or covered with snow) during the winter for protection, since the root systems require much less respiration and stop most growth.[9] For the winter the plants can be left alone or given some mulch to protect them in the colder parts of the United States.

Constructing a Small Cranberry Bog: This sounds like a massive excavation project with a bulldozer but it is actually simple to accomplish on a small scale when you understand that a "bog" is

Cranberries ripening.

Flowers with developing fruit.

Unripe berries amid foliage.

Cranberries ripening in fall.

anything that holds moisture throughout a growing season and allows the roots of moisture-loving plants to thrive in the right soil conditions. Cranberries have root systems that spread out and are shallow, so any low plastic container that contains moist soil but permits extra rainfall to drain away from the plants will allow the vines to thrive. Below is a simple way to create a small bog, but this example could be expanded into a bigger area using a large impermeable thick plastic sheet inside of a shallow hole.

Creating a Bowl Bog: Use any watertight plastic container that is big enough to allow several plants' roots systems to spread out—at least 8 inches (20.3 cm) deep. Dig a hole in the ground large enough to fit the entire container so that it sits flat on the bottom of the hole with the top of the container resting about ½ inch (1.3 cm) below the top of the planting area. Fill the container to the top with a three-part mixture of sand, garden soil, and an acidic amendment such as pine duff or peat moss; plant the vines into this soil and slowly add water until it completely saturates the soil up to the top of the container. Then use the original soil from the hole to pack around the perimeter of the container until it covers over and hides the edges of the bowl.

Hardiness: The hardiness range is dependent on the cultivar, but wild cranberry plants are rated hardy to zone 2 or −50°F (−46°C).

Fertilization and Growth: Cranberry is slow growing, so it is best to use several plants in one area. They grow by horizontal runners that spread out, and after many years they become crowded and tangled up with one another, which ultimately reduces the amount of sunlight that can reach the lower parts of the vines. An annual pruning back of some of the vigorous runners in early spring will add to the productivity of the plants and increase air circulation around their roots. Applying a mulch every two years in spring composed of an acidic medium such as pine needles or pine duff, mixed with sand, will help protect new shallow cranberry roots and introduce some acidic nutrients to the soil. The first year that cranberries are planted, they should be watered about once a week if the top area around the plants is dry to a depth of 2 inches (5.1 cm). Then after a year the plants should thrive if they receive consistent rain throughout the season, but may need watering during prolonged periods with no rain.

Cultivars: A century ago there were many cultivars of this fruit developed for different regions of the United States, with various ripening times and disease resistance.[10] While there are still many cultivars of cranberry used on commercial farms, these plants are not available at retail plant nurseries; only a few varieties are available to the gardener.

'Pilgrim': A hybrid cultivar introduced in 1961 that produces abundant crops of large, purplish red berries that ripen late in the season.

'Stevens': A vigorous and productive cultivar with deep red fruit that ripens midseason. It was introduced in 1950 and is tolerant of poor soil.

Related Species: The small cranberry (*V. oxycoccos*) is a European relative that is closely related.

Propagation: This plant is most easily propagated by digging up small runners of the vine with some roots attached to the bottom of the stems, then repotting them in acidic, sandy soil. The plants can also be propagated through cuttings potted up with a mixture of sand, compost, and peat moss for a few years until their root systems are larger.

Pests and Problems: If your plants are properly sited in the correct soil conditions, they can often be fairly pest-free. Cranberries can get a few insect pests and root diseases, but we have never had any problems. The worst damage to our plants was by voles that spent one winter destroying most of our vines; after this it took a few years for our plants to grow back. Unopened flower buds and young flowers can be damaged by late-spring frosts, which is one of the reasons commercial cranberry farmers flood their crops.

Elderberry

Sambucus canadensis, S. nigra ssp. *canadensis*

American Elderberry

> *If the medicinal properties of its leaves, bark and berries were fully known, I cannot tell what our countryman could ail for which he might not fetch a remedy from every hedge, either for sickness, or wounds.*
>
> —John Evelyn, 1664[1]

This is another plant that humans have used since the beginning of our civilization. Ancient cultures in Europe used various parts of elderberry plants for medicine, musical instruments, beverages and food, and even as a protection against witchcraft.[2] It should be noted that all the green parts of this plant are toxic because they contain cyanogenic glycosides; they should never be eaten. The berries should be picked only when fully ripe and their color dark purple to black. So although eating a few berries out of hand is okay, they are slightly toxic and eating too many will make you sick. More important, most people find that raw American elderberries have a slightly disagreeable flavor and taste much better when cooked. Because the fruits have lots of vitamins and minerals, they are a common addition to commercial organic cough syrups and drops, and many people (us included) make a traditional homemade cough syrup using the berries. Elderberries are a nutrient-rich fruit due to their high concentration of antioxidants.[3] Elderberry flowers are also an excellent traditional food flavoring, and Europeans have brewed alcoholic beverages such as cordials using both the elder flowers and berries.

American elder has a wide geographic distribution across large areas of the eastern United States, ranging from Nova Scotia to Florida, west to Texas and Minnesota, with scattered populations in mountain areas in western North America. This widespread distribution also underlies how tough elderberry is and its adaptability as a landscape plant.

In the past there was little interest in breeding improved elderberry varieties because they were so common as a roadside weed, but in 1997 the Elderberry Improvement Project was started at the University of Missouri to create improved fruiting cultivars suitable for the

Elderberry leaves and flowers.

Midwest.[4] In the last two decades, people have become more interested in this native medicinal and edible shrub, and there are now about a dozen varieties available in commerce.[5] We think this is a great, trouble-free combination of an edible, medicinal, and ornamental plant that should be more widely grown as a worthy addition to gardens.

Growth Difficulty Rating: 1. This plant is easy to grow with minor management for cane borers.

Taste Profile and Uses: Dark purple to black round berries are slightly less than ¼ inch (6 mm) in diameter, ripening from August through September, which are produced in big clusters, with larger-sized fruit in cultivated varieties. When cooked, the berries have a mild sweet flavor that is close to the taste of a not-so-sweet blackberry. When cooked in water with a sweetener and then strained, elderberry juice makes a superb nutritious fruit water. It is also great in pies, jelly, and sauces. Elderberry flowers are also used in a delicious traditional dessert fritter, when dipped in batter, fried to golden brown, and dusted with powdered sugar. As stated above, the flowers are used as a flavoring for wines and are a great flavor addition when making gooseberry jelly. Elderberry is also loved by wildlife—at least forty-two species of fruit-eating birds like the berries.[6]

Plant Description: A large suckering shrub or tree-like shrub growing from 6 to 15 feet (1.8–4.6 m) tall, with oppositely arranged compound leaves that are suggestive of ash (*Fraxinus* spp.) foliage. The 2- to 6-inch (5.1–15.2 cm) long, bright green leaves are narrowly ovate-shaped and have sharply serrated edges with leaflets that are formed into groups of five to eleven leaves, although normally seven. Older bushes survive by being stoloniferous (a horizontal stem that roots at its tip and produces new plants), so new shoots form branches as older stems die out after a number of years. The height of this species is variable, so there can be great differences in individual bushes.

Flowers: The plant produces decorative, large umbels (flower clusters) that can range in size from 6 to 10 inches (15.2–25.4 cm) across and are composed of small, white flowers with a pleasantly strong, musky scent. The flowers open in late spring to early summer, and mature bushes can be completely covered with blooms, making these shrubs an attractive addition to any landscape.

Pollination Requirements: Elderberry is self-fertile but will produce more fruit with another elderberry planted nearby.

Site and Soil Conditions: Elderberries are adaptable to many growing conditions, except for very dry areas. We have seen elderberry thriving in deep clay soil that is consistently moist; in the middle of forests in deep shade; in boggy conditions; and in open sites on the sides of the road in full sun. Muddy soil is one of the most difficult conditions for growing food. Because elderberry thrives in soggy, wet soil (but not constantly standing in water), it is a great choice for growing a food crop in a moist area that would kill most other fruiting plants. We have found that elderberry does fruit in shady conditions, but when planted in full sun with adequate moisture the plant produces plentiful amounts of berries.

Hardiness: The plant is hardy to zone 4 or −30°F (−34°C), but many sources list this shrub as hardy to zone 3 (−40°F/−40°C).

Fertilization and Growth: Though American elderberry thrives like a weed in our region in waste places with poor-quality soil, it will do best in consistently moist soil with some compost added as a mulch.

Cultivars: There are currently more than a dozen cultivars of American elderberry available in nurseries and through mail-order specialty plant catalogs.[7]

'Adams': A popular, vigorous-growing bush that reaches about 8 to 10 feet (2.4–3.1 m) in height, noted for its large-sized berries.

'Bob Gordon': A very high-yielding cultivar of elderberry, selected from wild plants found near Osceola, Missouri, in 1999.

'Nova': A slightly smaller-sized shrub, generally reaching to 6 feet (1.8 m), bred for abundant crops of berries.

'Wyldewood': A vigorous-growing bush with large-sized fruit that was introduced from the Elderberry Improvement Project in 2010.

'York': Bred for abundant crops of large-sized berries.

Related Species: The blue elderberry is native to the western United States (*S. nigra* ssp. *cerulea*), sometimes relegated to its own species: *S. cerulea*. It is an attractive plant featuring handsome blue-colored fruit, and reaches 8 to 12 feet (2.4–3.7 m) in height, taller plants may be found in the wild. The fruit is edible only when cooked.

European Elderberry: American elderberry (*S. canadensis*) is now believed to be a subspecies of European elderberry (*S. nigra*), so its new official nomenclature is *S. nigra* ssp. *canadensis*. We have tried to grow several cultivars of European elderberry over the years

Elderberry flowers.

Young fruit forming.

Elderberry fruit ripening.

Fully ripe berries.

and have killed over a dozen specimens, while a 'York' cultivar of American elderberry has done well for eighteen years as a neglected plant. In our gardens, European elders will grow for several years and then suddenly die without showing any apparent signs of distress, and we have no idea why. Although almost all the mail-order nurseries say that there are no serious pests to the European species, our environmental conditions seem to affect them in some way. With that said, we still feel that European elderberry may be a valuable edible landscaping plant and is worth trying out.

Propagation: Elderberry is one of the easiest plants to propagate. Gather seeds in the late summer, then plant them in pots and store over the winter. Hardwood cuttings that are 10 inches (25.4 cm) long taken in late fall or early winter from the previous season's growth will form new plants. Digging up suckering shoots from the base of the bush will generate new plants as well, particularly when you dig them up in spring before the buds of the leaves open.

In our environment these native plants are loved by birds, who eat the fruit and disperse the seeds, so we have plenty of "volunteer" plants that can be transplanted easily.

Pests and Problems: This shrub is fairly pest-free with the exception of a few insect predators in the eastern United States, such as the attractive elderberry borer beetle (*Desmocerus palliatus*), which we try to tolerate because they are so beautiful, with their cerulean blue and bright yellow-orange wing covers, so long as they do not get out of control. The beetles' presence on the plants is noticeable by the small holes on the stems. The beetles deposit their eggs into these holes. After the larvae chew through the pith of these stems, the foliage slowly wilts and dies. The affected stems can be cut to control the beetles if the problem spreads across too many branches of a bush. Another related beetle (*D. californicus dimorphus*) affects elderberries in California, but these beetles are currently on the Endangered Species List and care should be taken to allow some of them to live around your plants because of the importance of biodiversity!

There are also a few minor insect and powdery mildew issues that are usually not major problems for *Sambucus*. Honestly, in our gardens the biggest problem with ripening fruit is that they are often taken by catbirds.

European Quince

Cydonia oblonga

Quince

> *The quince, "the golden apple" of the ancients, once dedicated to deities, and looked upon as the emblem of love and happiness, for centuries the favorite pome, is now neglected and the least esteemed of commonly cultivated tree-fruits.*
>
> —U. P. HEDRICK, 1922[1]

The European quince is a fruit that has largely been forgotten by the modern world but played an important role in the beginnings of human civilization. Quince is considered to be one of the oldest cultivated fruits and believed to have been one of the trees in the biblical Garden of Eden. The fruit's botanical name *Cydonia* comes from the name *Cydon*, which was an ancient city on the island of Crete where the fruit grew abundantly. There is a long tradition of quince being mentioned in classical Greek literature that goes back to antiquity.

Quince is believed to have originated in the Near East and western Asia, but its cultivation as a food crop has spread across large portions of Europe.[2] Quince was brought into Britain with the French soldiers through the Norman conquest of England in 1066. England eventually acquired a taste for quince, and monarchs like King Edward I loved it so much that he had four trees planted in 1275 near the Tower of London. In England by the seventeenth century there were four named cultivars of these trees, one of them called 'Portingall' (also known as 'Portugal'), and over four hundred years later this variety is still sold in nurseries.[3]

America was introduced to quince by the Pilgrims, who brought the seeds from northern Europe, and it was planted in many historic orchards, such as Thomas Jefferson's Monticello property.[4] Quince became a useful nineteenth-century farmer's tree and was the first and only fruit source for commercial pectin production in America until the end of the nineteenth century.[5]

Today the center for quince fruit production is eastern Europe and Turkey, and this area of the world has introduced many of the improved, modern cultivars that are available in the mail-order nursery trade.[6] Because of its adaptability to different types of soil and its close biological relationship to pears, the quince has traditionally been used as a dwarfing rootstock for grafting pear varieties. Modern culture has viewed the European quince as an inferior fruit of little value because it is hard and cannot be eaten raw and must be cooked with a substantial amount of sweetener in order to bring out its culinary attributes. But we agree with the older European cultures that view this fruit as a spectacular and delicious one.

Growth Difficulty Rating: 1. European quince is easy to grow with minor management for cedar apple rust and fire blight.

Taste Profile and Uses: The intensely fragrant fruit is technically considered a pome, yellow to yellowish green, with an outer skin that has a waxy texture or is covered with a fine layer of fuzz. Quince comes in a variety of shapes and sizes depending on the cultivar, with shapes ranging from the fruit looking like a pear (pyriform) or an apple (round), growing from 3 to 5 inches (7.6–12.7 cm) in diameter. Before the fruit can be cooked, the fruit must be peeled, its many seeds removed, which is viewed by the modern world as too much work. To bring out the flavor, the fruit must be cooked until it turns a deep scarlet, which also releases a complex aromatic flavor that combines the tanginess of pineapple with a rich floral fragrance. When cooked with a sweetener, quince attains a flavor that is unlike any other fruit in the world. It is used for jelly, for paste, the Spanish *membrillo* (great with cheese), for sauces, and in an extensive variety of traditional Mediterranean recipes.

Plant Description: A slow-growing, small-sized deciduous tree reaching 10 to 12 feet (3.1–3.7 m) tall. The alternately arranged leaves are ovate to elliptical in shape, grow 2 to 4 inches (5.1–10.2 cm) long, and are a deep green color on top and pubescent (hairy) underneath.

Flowers: Attractive 1½- to 2-inch (3.8–5.1 cm) flowers bloom on the end of the current season's branchlets from April to May in large quantities, opening up after the leaves unfurl. The five petals of European quince flowers are a soft pink color with white centers, which gives them an attractive striped-pink appearance. Young quince trees are often completely covered in flowers.

Pollination Requirements: Quince trees are self-fertile and productive, though planting two trees together will increase fruit yields.

A 'Boyer' quince flower.

Beautiful pink-and-white flowers.

'Boyer' quince in bloom.

'Tekkes' Turkish quince.

'Kuganskaya' Baltic quince fruit.

Site and Soil Conditions: The fact that quince cultivation has spread across the world is a testament to the tree's adaptability: from hot, dry climates (Mediterranean) to cold, snowy climates (New England and northern Europe). The root systems of quince trees seem to do well in many types of soil conditions, but this is one of the few fruit trees that does well in heavy, moist soil.[7] While full sun is the best for fruit production, quince trees will grow in a half-shade planting site.

Hardiness: Quince is rated hardy to zone 5 (−20°F/−29°C), but is listed as surviving in zone 4, or −30°F (−34°C), by many mail-order fruit catalogs.

Fertilization and Growth: Slow-acting fertilizers like organic compost or leaf humus would benefit the tree. Avoid using any amendments that have excessive amounts of nitrogen because of the dangers of fire blight (see the "Pests and Problems" section).

Traditional Quince Cultivars: Currently there are over a dozen cultivars of quince available through specialty mail-order nurseries.

'Boyer': A productive old variety with small to medium-sized, yellow-green fruit that has a good, strong flavor.

'Portugal': An old European cultivar that produces large pear-shaped fruit of a good quality.

'Van Deman': An 1890s cultivar developed by Luther Burbank that bears good harvests of large-sized yellow fruit.

Modern Baltic Cultivars: Over the last decade Turkey and the Baltic region have introduced a group of new cultivars that have been selected for improved palatability. These quince cultivars are picked in the fall and stored for a few weeks, which increases the flavor and softens the flesh enough to be eaten fresh when cut into thin slices, but it is better cooked.

'Aromatnaya': A fairly disease-resistant cultivar from Russia that produces large amounts of 3-inch (7.6 cm) yellow fruit that is fragrant and good for cooking.

'Kuganskaya': A disease-resistant cultivar from the Caucasus Mountains with large yellow fruit that have a mild pineapple-like flavor and that can be eaten fresh off the tree but is also great cooked.

Related Species: Chinese tree quince (*Pseudocydonia sinensis*) is a beautiful small-sized tree that grows from 10 to 20 feet (3.1–6.1 m) high, with attractive viridian green and gray camouflage-mottled bark. The tree produces large fruit that is 5 to 7 inches (12.7–17.8 cm) long, but has less of the spicy floral flavor found in the European and flowering quince types. Though rated to be hardy to zone 5, or −20°F (−29°C), our tree was killed by a hard, cold winter, so we would only recommend this tree for zone 6 (−10°F/−23°C) or warmer climates.

Propagation: Hardwood cuttings taken in late fall to early winter will produce trees. Specific quince cultivars can be grafted onto quince rootstock.

Pests and Problems: Though a tough and mostly pest-free tree, quince has two serious problems: fire blight (a bacteria) and cedar apple rust (a fungus). Fire blight is the major problem that affects quince trees in many parts of North America. A naturally occurring bacteria (*Erwinia amylovora*) emerges on the tree as a black, shiny appearance, so that it looks like the leaves have been burned, and all the areas impacted by the bacteria wither and eventually die. The only way to deal with fire blight is to prune all the wood up to 1 foot (31 cm) below the affected regions of the tree, as well as cleaning the blades of the pruners with alcohol or bleach solution between each cut on the tree so that the disease is not spread by the pruners. The bacteria can be spread across all parts of the plant (leaves, stems, dead fruit) and should be cleared away and thrown out or burned. Rapid growth brought on by excess fertilization can also encourage the spread of fire blight, which is why fertilization with manure is not recommended for quince trees.

Cedar quince rust (*Gymnosporangium clavipes*) is a fungus that can spread as a disease only when host plants from the pome family (apples, quince, pear, and the like) are planted in proximity to members of the cedar genus (*Cedrus*). This fungus is spread by environmental factors, like wind and rain, and can travel for a great distance between trees, making this disease cycle difficult to control in forest settings. It is not always possible to select rust-resistant cultivars, but there are some ways you can reduce the impact of this problem. The disease may be kept in check by removing the infected dead fruit and branches from the tree through judicious pruning, sterilizing the pruners between cuts in the manner stated in the description of fire blight above. The areas of the tree damaged by rust are easiest to spot in the fall after the foliage has dropped off the tree. Make sure you dispose of the damaged tree parts and fallen leaves, which may also be infected with rust, away from your normal compost pile or you can continue to spread the disease to other plants through your use of compost.

Flowering Quince

Chaenomeles spp.

Ornamental Quince

> *The flowering quinces are as difficult to classify as they are beautiful to look upon. Even as their flowers are borne upon twisted thorny branches so the whole question of their nomenclature forms a spiny and painful thicket which must be penetrated before any precise discussion of garden-worthy varieties can be undertaken.*
>
> —EDGAR ANDERSON, 1935[1]

Chaenomeles bushes are generally known as old-fashioned garden plants that are harbingers of spring because they produce abundant displays of bright, colorful flowers that appear in late winter as the plant's leaves begin to open. Most people refer to these shrubs as flowering quince, but the precise identification of individual species is complex, and there is a long history of confusion among both plant taxonomists and nursery retailers. In 1784 Sir Joseph Banks, the director of the Royal Botanic Gardens at Kew, introduced the first Japanese quince to England, which was mistakenly identified as a species of pear.[2] That was only the start of the misunderstanding of these plants, which were then identified as quince and renamed *Cydonia japonica*. Then in 1822 the British botanist John Lindley created the genus *Chaenomeles* (flowering quince) and separated it from *Cydonia* (tree quince) and renamed it *Chaenomeles japonica*.[3] As it turned out, this particular species was actually from China but had been cultivated in Japan for many years.

The history of flowering quince only gets more perplexing as a few other species were introduced into cultivation and given accurate (or inaccurate) species names, then hybridized to produce a bewildering array of new plants. By the 1960s there were actually over five hundred known cultivars of Japanese quince.[4]

Chaenomeles shrubs have been in American gardens for over a hundred years, and at one time it was one of the most popular decorative garden and hedge plants in this country, especially in the central and southern states, where old, unruly specimens can still be found.[5]

Few gardeners are aware of the fact that flowering quince can be grown for their fruit. Modern plant breeders decided that the fruit on chaenomeles bushes is an unwanted benefit, and they began breeding double-flowered cultivars that are sterile, producing no fruit. Almost one hundred years ago U. P. Hedrick, one of America's great fruit experts, noted that the different species of *Chaenomeles* offered the plant breeder a good opportunity for hybridizing better-quality fruit.[6] We hope that, in the future, plant breeders do indeed create plants with larger fruit and smaller seeds, but in the meantime we think that the jelly produced from flowering quince is incredibly delicious.

Growth Difficulty Rating: 1. Easy to grow in most types of soil with very few problems except for some cedar rust issues. After a few seasons the bushes are virtually indestructible.

Taste Profile and Uses: The flowering quince is a pome, with rounded fruits that are yellow to yellowish green and about 1½ to 2 inches (3.8–5.1 cm) in size. It is not a fruit for fresh eating. It must be cooked with a sweetener to make it palatable, and it requires a substantial amount of time for peeling, coring, and removing the numerous seeds. After reading the instructions above, some readers may wonder if the fruits are even worth the effort, but quince jelly tastes like an extraordinary combination of sweetened cranberry and lemon, and is a beautiful deep scarlet-red color. The best way to make it is to cover the peeled and cored fruit with water and cook the mixture on the stove until the fruit

The early-spring flowers of 'Texas Scarlett'.

'Rubra' in full bloom.

Flower buds opening.

Fruit developing.

can be pierced with a fork. If you separate the water from the cooked fruit, both can be used as a food product. The water contains pectin, which can be cooked down to make a superb, piquant scarlet-colored jelly that has a floral taste rich with lemony notes. The strained fruit pulp has a sharp, aromatic taste and is used to make *membrillo* (fruit paste) or can be added to cooked fruits such as apples to make a floral-tasting applesauce.

Plant Description: Chaenomeles are multi-stemmed deciduous shrubs with branches that can be thorny. The simple leaves have finely serrated edges, are a rich green color, and are alternately arranged on the branch. The bushes range in size from 2 to 10 feet (0.6–3.1 m) tall depending on the species and cultivar, but most of the cultivated plants grow about 5 to 8 feet (1.5–2.4 m) tall and wide.

Flowering quince bushes often have an unkempt look when left unpruned that has been referred to in garden literature as having a growing habit like a "garbage can" since old leaves and even the occasional piece of garbage can collect in its twiggy mass of branches.[7] We think this is a bit harsh, but the shrubs do need to be pruned occasionally to keep them looking tidy.

Because of the wide range of flowering quince varieties, there are bushes to fit almost any garden space—that is, if you can actually find one! It has become more difficult because the nursery trade is currently not very interested in selling many of the cultivars that produce fruit.

The beautiful early-spring flowers of 'Tanechka'.

Chaenomeles japonica produces some of the shrub varieties that have a shorter stature with spreading branches and a tendency to grow horizontally, reaching 3 to 4 feet (0.9–1.2 m). *Chaenomeles × superba* is slightly larger, growing to about 4 to 5 feet (1.2–1.5 m); *C. speciosa* has branches that grow taller and slowly spread out with an ultimate height between 6 and 10 feet (1.8–3.1 m).[8]

Flowers: The flowers open up just after the leaf buds have cracked open in early spring, and can produce a beautiful display of well-needed color at a time that most landscapes are gray and still waking up from winter. The blossoms are generally pink, red, or orange, but there are cultivars that come in a range of colors such as white, yellow, and variations of pink with white. Flowers grow from 1 to 2 inches (2.5–5.1 cm) across, depending on the species and cultivar. An occasional second flowering may occur during abnormally warm fall weather.

Pollination Requirements: Flowering quince blossoms are self-sterile and require the pollen from another quince bush in order for the plants to set fruit.

Site and Soil Conditions: Chaenomeles are fairly adaptable and grows in a wide range of conditions as long as the soil is well drained. The shrubs can be grown in both part shade to full sun, with the most abundant flowers produced in full sun. Flowering quince are considered to be fairly drought-tolerant after getting established in a planting site for a few years.

Hardiness: There is some difference of opinion as to the cold tolerance for these plants, with some sources listing them as hardy to zone 5, −20°F (−29°C), and some rating them at zone 4, or −30°F (−34°C). In the areas of the United States that are in the coldest parts of zone 4, these plants may be damaged or killed during the worst winter conditions.[9]

Fertilization and Growth: Some compost added to the soil around the roots would encourage the shrubs to produce more flowers and grow more vigorously. Chlorosis (yellowing leaves) can occur if quinces are planted in soil that is too acidic, such as under established groups of pine and oak trees.[10]

Cultivars: There were literally hundreds of cultivars of flowering quince in the middle of the twentieth century, and while this plant has fallen out of favor, there are still some retail nurseries and specialty mail-order catalogs that carry a few varieties. At the time of this publication, the dominant popular flowering quince bushes available in most plant nurseries are the new introductions of the patented 'Double Take' series, and while these are beautiful double-flowering varieties, they are all sterile and will not produce fruit. Below is a list of a few cultivars that are productive fruiting plants that also have beautiful flowers.

- **'Crimson and Gold' (*C.* × *superba*):** Magnificent fire-engine-red flowers with golden anthers on a shrub that tends to sucker and grow horizontally, from 3 to 5 feet (0.9–1.5 m) high.
- **'Jet Trail' (*C.* × *superba*):** A compact shrub that grows 3 to 4 feet (0.9–1.2 m) high and produces abundant displays of pure white flowers.
- **'Tanechka' (*C. japonica*):** A Ukrainian cultivar with attractive scarlet-pink flowers, that grows 5 to 6 feet (1.5–1.8 m) high and produces large crops of fruit.
- **'Toyo-Nishiki' (*C. japonica*):** A distinctive Japanese cultivar with pink-, white-, and scarlet-colored flowers on the same shrub that grows 4 to 7 feet (1.2–2.1 m) in height.

Related Species: *Chaenomeles* is also related to Cathay quince (*C. cathayensis*) from China, which grows as a tree from 9 to 12 feet (2.7–3.7 m) in height. There is also another attractive tree with exfoliating, highly decorative bark commonly called Chinese quince (*Pseudocydonia sinensis*). Both of these species produce edible fruits that need to be cooked.

Propagation: Flowering quince is easily propagated from hardwood cuttings that are taken in late winter to early spring before the first leaf buds crack open, and then potted up in soil for three growing seasons until they have a good root system for replanting in a new location.

Pests and Problems: We have not had any major issues with this plant, although cedar rust (*Gymnosporangium clavipes*) affects some of the fruit each season. We gather the diseased fruits and burn them or dispose of them in the garbage to prevent the rust developing into a bigger problem. Pruning some of the older branches on old, established bushes will allow more air circulation and sunlight, which will lessen the impact of fungal diseases.

Scale insects can be a problem in certain parts of the United States, but they do not usually cause a significant amount of damage on most shrubs. The fruit is so hard that most animals leave them alone; a few may try chewing some but usually abandon them because of the sour taste.

Goji

Lycium barbarum, L. chinense

Chinese Boxthorn, Duke of Argyll's Tea-Treey, Matrimony Vine, Wolfberry

The Goji Well

A cool well beside the monk's house,
A clear spring feeds the well and water has great powers.
Emerald green leaves grow on the wall,
The deep red berries shine like copper,
The flourishing branch like a walking stick,
The old root in a dog's shape signals good fortune.
The Goji nourishes body and spirit.
Drink of the well and enjoy life.

—LIU YUXI, Tang dynasty, circa 800 CE[1]

Goji (or wolfberry) is a vine disguised as a bush, or a bush with vine-like tendencies and a messy habit. Either way you look at it the goji is a big sprawling plant that has long been revered as a national treasure in China because the berry it produces is believed to be one of the most nutritious foods on Earth.[2] The native range of goji shrubs is

Flowers, buds, and developing fruits.

Jewel-like fruit.

Fruits and flower.

from southeastern Europe to northwestern China, and the fruit's cultivation in China dates back to 1000 to 1400 CE.[3]

Though there are many species of shrubs and vine-like plants in the genus *Lycium*, the two plants that are actually called goji are the *L. barbarum* and *L. chinense*. The two are very similar in growth habit, but *L. barbarum* is the species most commonly encountered in nurseries. It is larger in all its attributes, from its size to its flowers and fruits, with the added benefit of surviving colder winter temperatures.[4] The fruits of both species have been used in traditional Chinese medicine for restoring energy to the body and are believed to be a cure for a wide range of maladies from skin rashes and eyesight problems to diabetes, and there are well over fifty scientific research papers from China, Europe, and the United States dedicated to the nutritional benefits of this humble but revered berry.[5]

Lycium barbarum is primarily found as a wild plant in the Ningxia Hui region, a small autonomous area located in north-central China renowned for having the largest-scale agricultural production of the fruit for medical uses. It's also famed for producing the highest-quality berries.[6]

Goji was only introduced to American consumers about a decade ago. It has slowly become popular because it is considered a nutritional superfood and is commonly sold in natural food stores as a dried fruit or as an ingredient in trail mixes or granola. The plants have a messy habit, but if they are planted in the proper site with enough room to grow, this low-maintenance shrub's sprawling arched branches with dangling red berries create an attractive ornamental feature.

Growth Difficulty Rating: 2. Goji is easy to grow but will ultimately require some pruning maintenance. You'll also need to train the bushes against a fence or trellis to keep them in a manageable shape and prevent them from spreading out of control.

Taste Profile and Uses: Ripe fruits range in color from dark orange to scarlet red with an oval shape and are highly decorative. The vining shrubs can set fruit for up to two months and produce ripe berries over a long period of time. Gojis acquire the best flavor when they are completely ripe with skins that are just starting to wrinkle before harvesting. The berry is generally ½ to ¾ inch (1.3–1.9 cm) long with soft pulpy flesh and multiple tiny, edible seeds. The fresh berries have a watery, astringent tomato taste, and the best way to use goji as a food crop is to dry the berries, which concentrates and intensifies their flavors. When dried the fruit takes on a burgundy color and the flavor is sweeter with a slight tang that tastes like a combination of licorice and cranberry. The berry is now often used in many trail mixes because it is nutrient-dense; it contains higher levels of antioxidants than even blueberries, along with dozens of healthy minerals and very high levels of carotenoids, which are beneficial to eyesight.[7]

Although mostly eaten dried, the fruit is also used in traditional Chinese recipes, and we have had visitors to the garden recall childhood memories of eating chicken soup with fresh goji berries. The young leaves of the plants are also edible and harvested fresh as healthy greens for cooking or as a tea.[8]

Plant Description: Goji is a deciduous woody shrub that grows in a dense, thick form with a sprawling growth habit. It produces long, slender branches that arch in in a similar fashion to a blackberry cane. *Lycium barbarum* can grow 6 to 8 feet (1.8–2.4 m) tall and spreads wider than that because the branches have a tendency to flop sideways like a vine. *Lycium chinense* grows between 3 and 6 feet (0.9–1.8 m) tall. Both types spread out with root suckers that produce new growth away from the original root system and form bramble-like goji colonies.

Young branches can grow many feet in a single season, with thorns lightly distributed along them. The elongated lance-shaped foliage ranges from ¾ to 2½ inches (1.9–6.4 cm) long, is bright green and alternately arranged, and produced in bundles of up to three leaves that are spaced apart along the stems of the plant.

The life span of the bushes is confusing with some sources stating that gojis are considered to be short-lived, from about seven to ten years, but because they grow a wide, fibrous root system, they can produce new plants by resprouting new stems from the roots, which enables the plants to have longer lives. Other sources suggest that the shrubs can live over a hundred years.[9]

Flowers: The tiny buds of the unopened flowers hang in the leaf axils like little teardrops along the branches and open up as small ¾- to 1-inch (1.9–2.5 cm) wide violet flowers with five petals and a prominent stigma. After pollination the fruit forms quickly, and often both the flowers and fruits appear on the same branch at the same time.

Pollination Requirements: Flowers are bisexual and self-fertile, but better pollination occurs with more fruit-set if more than one shrub is planted.[10] Because the fruits are teeny, our advice is to plant two or more shrubs to have a substantial harvest.

Site and Soil Conditions: Goji bushes prefer full sun but can take some morning shade as long as they have well-drained soil. They are adaptable and can thrive in rich, fertile soil, but in the wild they grow in poor-quality soil conditions and do well on most sites, except in soil that is soggy and wet. Once established the plants can take arid conditions and have been used for erosion control.[11] The plants in our arboretum have thrived in soil amended with rich organic compost that is well drained. Because of goji's tendency to be floppy it produces many root suckers that ultimately form into a large edible hedge.

Hardiness: The goji commonly sold in garden nursery commerce is *L. barbarum* and is rated to a cold hardiness of zone 5, or −20°F (−29°C), although some sources list particular cultivars of that species as being tolerant of colder temperatures. *Lycium chinense* is rated to be hardy to zone 6, or −10°F (−23°C).

Fertilization and Growth: Goji shrubs are precocious, and a moderate-sized bush will often produce a small amount of fruit the year after it's planted. Goji fruits on new wood each year, so you can prune back the plants in early spring when they are still dormant to keep them more manageable in a garden setting. Because mature shrubs produce large quantities of arched branches, carefully pruning away damaged, old, or tangled branches every few years will encourage new growth and allow for good air circulation.

Training, Staking, and Using Goji Bushes on a Wall: In order to deal with the goji plant's floppy growth habit, many gardeners

Goji spilling over stone wall.

locate the bushes near a garden fence, training sections of the plant up the fence with wire twist-ties in order to keep the fruit accessible for picking. A trellis, or strong wooden stakes, can also be used as a support system to tie up several thick branches of a large bush. For many years we pruned our plants into a shorter, manageable size; then we decided to embrace the goji's wild growth habit and replanted our bushes on the side of a small stone wall embankment so the plant attractively spills over the sides of the wall.

Cultivars: Jim Gilbert and Lorraine Gardner of One Green World Nursery traveled to Ningxia in 2005 and returned with one of the leading varieties of Chinese goji, 'Ning-qi No. 1', which they began selling as Crimson Star. This cultivar is now one of the predominant varieties sold in the US.[12] Since that time the goji has become more popular as a fruiting shrub and several more cultivars have been introduced. All the plants listed below are varieties of *L. barbarum*.

'Crimson Star': A trademarked cultivar that is popular in northern China for producing large crops of flavorful fruit.
'Golden Goji Berry': A productive yellow-fruited cultivar of goji with fruit that contains more juice and has a milder flavor.
'Phoenix Tears': The other dominant cultivar, it produces large harvests of berries in an elongated shape, with leaves that are rounded at the base.

Related Species: Goji is a member of the giant Solanaceae family (the nightshade family) and is distantly related to many better-known produce staples such as eggplants, peppers, and tomatoes. Black goji berry (*L. ruthenicum*) are fruiting bushes native to the Himalayan mountains and are also a healthy food used like the other species of goji.

Propagation: Seeds and cuttings will both produce new plants, but goji is easily propagated by layering the tips of branches, or dividing up sections of the shrub. As the plant matures, you can dig up the suckers with some roots attached to the ends when the plant is dormant. These can be potted up or planted directly into another garden location, if the root systems are large enough.

Pests and Problems: Our goji plants have not been bothered by any major diseases. Birds and other wildlife do occasionally eat the berries.

Gooseberry

Ribes grossularia, R. hirtellum, R. uva-crispa

The gooseberry is one of the lesser lights in the pomological firmament of the United States. It is apparently prized and appreciated less than almost any other fruit.

—FRED W. CARD, 1914[1]

Northern Europe, especially England, has had a long love affair with gooseberries, with the noted nineteenth-century English pomologist George Lindley listing 772 varieties of cultivated gooseberries in 1831. On the other hand gooseberries cultivated in the United States received very little attention by comparison, with one of the first crosses being recorded as 'Houghton' in 1833, and although it was said to possess decent flavor, the fruit was small by European standards. Breeders at that time were more interested in other fruits like blackberries and grapes, so little effort was made toward domesticating the American gooseberry or adapting the European types.[2]

The lack of popularity of gooseberries was also not helped when the growing and importation of all *Ribes* species was banned in the early nineteenth century because gooseberries (like blackcurrants) can spread white pine blister rust (see the "Blackcurrant" chapter). Even though many gooseberry plants are not as susceptible to spreading this disease and the federal ban on *Ribes* was lifted in 1966, the incorrect view that all *Ribes* fruits cause this disease has stuck to the fruit.[3]

Many of the early best-flavored cultivars introduced from Europe failed under the humidity and heat of eastern North American growing conditions; worse, European gooseberry cultivars were often killed off by a native species of fungus.[4] By the 1920s when the pomologist U. P. Hedrick tallied up the number of cultivars that were actually surviving and being grown in North America, it came to about twenty varieties,[5] which is very similar to how many are available to the home grower today, one hundred years later!

Many people today are unfamiliar with these fruits because America has never fully embraced the gooseberry, most likely because the bushes are often armed with sharp thorns, or they have only tasted an unripened one. This is a shame because the taste of a fully ripened gooseberry from a good cultivar can rival any of the popular dessert fruits grown in America. And even with the limited gooseberry choices cultivated today, there are a variety of flavors. While the United States may never experience the extraordinary diversity of this fruit that was available in nineteenth-century England, recent years have seen gooseberry cultivation begin to spread in the United States, with a majority of today's cultivars being crosses of the European gooseberry (*Ribes grossularia, R. uva-crispa*) bred with the American gooseberry (*R. hirtellum*) to give them disease resistance.

Growth Difficulty Rating: 1. Easy. We grow over a dozen different cultivars of these fruits and every year a few of our bushes get hit by currant sawfly, although the plants still bear gooseberries.

Taste Profile and Uses: If you taste a ripe gooseberry right off the bush, you'll immediately understand why this fruit is worth growing. The exterior skin of a gooseberry can range in texture from smooth and translucent or opaque, to slightly fuzzy or hairy, and there is a wide range of colors including golden yellow, green, red, purple, and black. They can vary in size from jumbo cultivars that are around the size of a cherry tomato, to fruit that is smaller than a blueberry. The round or slightly oblong berries have a pale, soft interior with fifteen to thirty tiny edible seeds.

Gooseberries typically ripen over a period of four to six weeks and require several pickings to harvest all the fruit. Although the fruit may grow to its proper size early on, the gooseberry is one of those fruits that stay sour for an extended period of time, then acquire a sweet flavor seemingly overnight. When harvesting the fruit, wait until they are dead ripe and the sugars reach their maximum point. Often this will be visible when the gooseberry acquires a blush of a darker color. If you are growing some of the darker-colored cultivars, you will need to taste a few at a time to see if they are ripe or not. Their distinct tart flavor makes them incredibly delicious eaten out of hand, as well as in fruit sauces, pies, and jellies.

Plant Description: Gooseberries are a deciduous, multi-trunked shrub with woody branches that are often armed with sharp spines, which grow from the axils of leaves, although there are a few cultivars that are nearly thornless. The alternately arranged leaves are 3⁄4 to 2½ inches (1.9–6.4 cm) in size and have three to five lobes with rounded tips. American cultivars can have fine hairs on the foliage. Depending on the particular cultivar, plants can grow from 2 to 6 feet (0.6–1.8 m) in height and can live up to 30 years.

Flowers: Gooseberries produce inconspicuous bell-shaped, greenish yellow to whitish pink blossoms depending on the cultivar; these are "perfect" (bisexual) and open in spring. The flowers grow either individually or arranged in groups of two or three on one-year-old wood as well as on spurs of older branches.

Pollination Requirements: These bushes are self-fertile but will produce more fruit with another plant nearby. Gooseberries bloom in early spring and is pollinated by insects and the wind.

Site and Soil Conditions: Gooseberries will survive in a diverse range of soils and conditions, but have a preference for consistently moist soil with some fertility. They are found in the wild in the shade of forests in northern latitudes and benefit from a heavy mulch to keep their roots cool in conditions of greater sun exposure. We grow gooseberries in ten to twelve hours of full sun, but these plants can do well with less sun exposure. In our gardens, even with heavy mulching on their root systems, the plants tend to look scorched, and most of the foliage falls off by midsummer, but the bushes produce large quantities of fruit each year and grow back vigorously each spring.

Hardiness: These plants are hardy to zone 4, or −30°F (−34°C), but some cultivars of gooseberry are hardy to zone 3 (−40°F/−40°C).

Fertilization and Growth: Gooseberries benefit from an annual application of well-aged manure or compost in early spring before the leaves bud out. American gooseberries can handle more sun than their European relatives, but the leaves get sunburned and fall off during a hot, dry summer—which is why we replenish

Gooseberry shrub.

Gooseberry flowers.

'Invicta' fruit.

A bountiful harvest of 'Tixia'.

them with another layer of mulch after the plants have fruited. This amendment to the soil will help keep the root systems from drying out. Gooseberries often lack the mineral potassium, so they benefit from a light annual addition of an organic amendment rich in potassium, such as ashes from a fireplace that contain the mineral potash.[6]

Cultivars: Modern gooseberry cultivars are mostly derived from European gooseberry and the American gooseberry. We divide gooseberries into two types: fresh-eating cultivars that are consumed right off the bush because of their sweet flavor; and the varietals that are tart-sweet and best for cooking. Below we have listed a few of the better-quality varieties.

Sweet Types:

'Invicta' (*R. uva-crispa*): A very productive, disease-resistant European cultivar that produces sweet, high-quality, green-colored fruit, introduced by the Malling Fruit Research Station in England.

'Jahn's Prairie' (*R. uva-crispa*): A 2- to 3-foot (61–91 cm) high bush with few to no thorns, discovered in the wild in Alberta, Canada, which produces large-sized sweet fruit with a thick skin.

'Poorman' (*R. hirtellum × grossularia*): A disease-resistant, productive cultivar developed in Utah around 1890, with abundant crops of high-quality fruit.

Sweet-Tart Types:

'Hinnonmaki Yellow' (*R. uva-crispa*): A sweet-tart cultivar from Finland with a delicious apricot-like flavor.

'Jewel' (*R. uva-crispa*): A Polish cultivar that produces an abundant amount of tangy, aromatic fruit with a red-orange color.

Related Species: All currants (*Ribes* spp.) are close relatives of the gooseberry.

Propagation: Gooseberries are one of the easiest plants to propagate. The traditional method is to take cuttings in late winter to early spring, but this must be done before the buds open. Choose sections of the stems that are ¼ inch (6 mm) in diameter and 7 to 10 inches (17.8–25.4 cm) long, near the top of the shrub with several healthy buds. Place the stem cuttings in soil with most of the buds covered up, leaving two buds exposed. Layering attached branches into the soil also works to produce new plants, and these layered stems can be dug up and separated from the mother plant in two to three years.

Training and Pruning: Although gooseberry bushes live a long time, the stems start losing their productive fruiting period after four years of growth. After a bush is six to eight years old, the best way to remedy this loss in fertility is by selectively cutting away older, unproductive stems to allow new shoots to grow around the bush. Different cultivars seem to vary in terms of how vigorously they grow. Some produce robust growth with lots of stems, and some are stingy and concentrate their energy into a few stems. So care should be taken in how much you cut off the plant.

When a gooseberry bush gets to a substantial size, it can become floppy, which means that the stems can bend outward toward the ground. When fruit is concentrated near the ends of branches that are on or near the ground, the fruit can get muddy and are more likely to be eaten by animals. There are several ways to deal with this problem. You can plant the bush near a garden fence, which allows you to train the branches on the fence to make the fruit more convenient to harvest. Or you can keep a bush upright by attaching several of the main branches to wooden stakes placed into the soil nearby.

Pests and Problems: These plants are targeted by a few pests such as aphids, currant borers, and gooseberry sawfly. These insects eat the foliage and can cause significant damage to the leaves. The gooseberry sawfly (*Nematus ribesii*) is usually the most significant pest for growers in the eastern part of the United States. Female sawflies lay eggs on the underside of leaves, low down in the center of the bush, so the young larvae go unnoticed until they have eaten their way upward and outward through the bush, devouring the leaves as they go. This becomes noticeable with the appearance of defoliated plant stems and insect frass (poop). The pest can be dealt with by removing larvae by hand or an application of an organic pesticide. Pay particular attention to the edges and undersides of the leaves where the light-green-colored larvae hide.

European varieties of gooseberry can get mildew (*Sphaerotheca mors-uvae*), which causes a powdery, gray-and-white fungus to appear on leaves and stems. The mildew may also appear on fruit, causing problems with ripening.[7]

Goumi

Elaeagnus multiflora

Cherry Silverberry

> *William Falconer . . . says that it [Goumi] is cooked and used as a sauce with meat, especially chicken, and "it is one of the most delicious sauces that ever tickled the human palate." Others do not speak so favorably of it . . .*
>
> —Fred W. Card, 1914[1]

Goumi is an adaptable, unfussy shrub that is native to the far eastern side of Russia, China, and Japan, where the fruits are eaten fresh and utilized for various culinary uses. It was introduced into America in 1889 by the Ellwanger & Barry Nursery in Rochester, New York, which originally sold the plants as an ornamental garden shrub.[2] The nurserymen said nothing about the edibility of goumi but touted the plant as a handsome ornamental bush due to its abundant displays of attractive scarlet fruit, which are indeed magnificent to behold. While goumi is a traditional fruit utilized in its native lands for its sour-cherry-like flavor, it was not until the late 1980s that a few mail-order nurseries in the United States specializing in unusual fruits began offering a couple of cultivars for sale. The most popular of these was a large-fruited cultivar called 'Sweet Scarlet' that originated at the Hryshko National Botanical Garden in Kyiv, Ukraine.[3]

It is unfortunate that goumi is not more common with homeowners looking for a carefree shrub that is both decorative and edible, because the plant has so many positive attributes. It is a low-maintenance plant that is pest free and very pollinator friendly, one that consistently bears heavy crops of tart-sweet, sour-cherry-tasting fruits that look like little gems.

Almost twenty years ago we planted a single goumi specimen in our garden, and within five years the plant began producing hundreds of berries. A single mature bush can provide thousands of berries for fruit preserves and fresh eating while also offering plenty of fruit for visiting birds.

Growth Difficulty Rating: 1. Goumi plants are easy to grow and adaptable to many growing sites.

Taste Profile and Uses: The oval-shaped drupe-like fruits are scarlet-red and covered with attractive tiny, metallic scales. The fruit is about ⅜ inch to ½ inches (1–1.3 cm) long with one edible seed; individual fruits ripen sporadically over a period of two to three weeks.

Goumi can be eaten out of hand but needs to be fully ripe or the fruit will have an astringent, unpalatable taste. Because the color of the fruit can be a misleading indicator of ripeness, we find that the best way to determine if they're ready is by "tickling" them with our fingers. This loosens mature fruits and allows them to easily fall off the long pedicles (fruit stems) into a bowl. A mature bush can be completely laden with fruit, which can persist on a shrub for up to a month. Dead-ripe fruit has the tart flavor of a sweet sour cherry or Cornelian cherry combined with some of the tangy qualities of a cranberry. Some people relish the fruits while others might find that they have an unusual taste better reserved for fruit sauces.

In addition to eating them fresh off the bush, goumi is great for making jams, jellies, sauces, and fruit leather. The seeds from the fruit can accumulate when making sauces and jelly, and we have found that the best way to deal with this problem is to cook down the fruits in some water until they are soft so they can be separated by using a ricer. The berries also have nutritional benefits with very high concentrations of lycopene, which is a powerful antioxidant.[4] Not only are the fruits tasty, but scientific research

The fragrant flowers.

Ripening fruit.

The decorative fruit.

The tiny metallic scales on the ripe fruit.

has indicated that they are rich in vitamins A and E, minerals, flavonoids, and other essential fatty acids.[5]

Plant Description: Goumi is a deciduous, multibranched shrub whose stems are slightly thorny; it grows 6 to 10 feet (1.8–3.1 m) in height. The plants have attractive, alternately arranged, bright green, 1½- to 2½-inch (3.8–6.4 cm), elliptic to ovate-shaped leaves that are covered with tiny metallic brown scales—a distinctive feature of the majority of plants in the *Elaeagnus* genus. The undersides of the leaves are a lighter color with metallic dots, and the young branches on the shrubs are a light chocolate color that also feature those attractive metallic scales, further adding to the overall ornamental attributes of the shrub.

Flowers: The small ⅝-inch (1.6 cm), pendulous, creamy-white flowers are generally borne singly in the axils of the leaves and hang in loose clusters in early spring. The flowers are apetulous (having no petals) with four sepals joined at the base of the blossom in the form of a tube, and both the sepals and tube are scaly with little golden dots all over them. The numerous flowers fill the air with an intense sweet fragrance, similar to the scent of lilacs, that can be smelled from a distance. The pollen from the flowers is loved by bees, and our shrubs literally hum with the sounds of insects.

Pollination Requirements: Goumi are self-fertile and productive, although many written sources on this plant suggest that you need two specimens for good fruit-set. In our experience of growing this plant over a long period of time, this is definitely not the case. Our two goumi bushes consistently produce large crops of fruit each year and they are planted 800 feet (244 m) apart.

Site and Soil Conditions: Our first bush was planted eighteen years ago in a location that receives six hours of sunlight in infertile soil with a heavy clay content. Our second goumi plant was a layering of the first plant and put in the ground four years later; it's sited in an area with up to ten hours of sun on land that has a gradual slope and is drier. Both plants are highly productive despite their differences in light exposure and soil conditions.

As long as goumi gets regular seasonal rainfall and is not constantly sitting in swampy soil conditions, it is a very adaptable shrub. Goumi takes nitrogen out of the air and puts it into the ground through a symbiotic relationship with the bacteria that live in their roots, which improves the fertility of the soils around them. This allows the plant to thrive in poor-quality soils, and also shares the nitrogen with other plants' roots that grow around them, which is why it is touted as a useful companion plant in orchards. Permaculture growers who grow fruit trees can utilize this as a nurse shrub in forestry plantings.[6]

Hardiness: Goumi has a cold hardiness rating to zone 4, or −30°F (−34°C).

Fertilization and Growth: Because its root systems can capture atmospheric nitrogen and fertilize the soil around them, goumi can grow in many types of soil conditions. Many members of this genus have survived in the difficult salty soils of maritime areas, so it is highly probable that goumi could also succeed in these areas.[7]

Cultivars: The improved cultivars of goumi were selected for fruit that is larger and more flavorful than the wild species form of the plant.

'Red Gem': A recent introduction by the Hryshko National Botanical Garden that originated in Russia, selected for improved fruit quality and productivity.

'Sweet Scarlet': A Ukrainian cultivar bred for fruit production and improved fruit quality.

Spring flowers on a goumi bush.

Related Species: Silverberry (*E. commutata*) and buffaloberry (*Shepherdia argentea*) are closely related. In many parts of the Northeast, the goumi's infamous naughty sister species autumn olive (*E. umbellata*) is considered a noxious, invasive shrub and has literally taken over whole swaths of land and wiped out native species.[8] Thankfully, our established goumi specimens have produced almost no viable seedlings, despite a long history of ripening literally thousands of berries. This plant could be a great alternative for anyone who loves the taste of autumn olive fruit but does not want to see their property overrun with a thousand autumn olive bushes.

Propagation: When our goumi bush started producing fruit, we spit out the edible seeds and planted them near the mother plant but surprisingly found that baby seedlings almost never sprouted. Some sources say the plant can be difficult to grow by seed, which can often take two years or more to germinate.[9] Thankfully it is a very easy shrub to propagate by ground-layering a young branch, air-layering branches, or planting pencil-sized cuttings in soil in late fall to early spring.

Pests and Problems: Our plants have not been bothered by any diseases. We often have bird nests in our bushes, and they will feast on the ripening fruits, but when the goumi are loaded with berries, there is so much to eat that we can share them with the wildlife and still have enough of a crop left over to make some jelly!

Invasive Plant Warning: As noted previously, in our gardens the individual goumi bushes are planted 800 feet (244 m) apart and have produced less than seven seedling plants in about twenty years. There are some sources that list this plant as being able to self-seed and therefore potentially problematic or mildly invasive in some growing conditions.[10] The discrepancy between our gardens not having invasive seedlings and other planting sites where goumi is reseeding itself may be the difference between a single bush planted alone in a landscape site as opposed to several bushes being planted together, so this should be taken into consideration when planting this useful edible plant.

Grapes

Vitis labrusca, V. vinifera

> *The common northern species,* Vitis labrusca, *produced good yields of large fruits but it was singularly unpromising from the standpoint of taste. Its common name, the "fox grape," referred to a distinctive odor that was linked with varying degrees of specificity to anal secretions.*
>
> —PHILIP J. PAULY, 2004[1]

The epigraph from a pomological historian highlights the early American immigrant's views on the supposed mammalian "bouquet" of wild American grapes. We have actually tasted delicious wild native grapes, but these were too strong for the fickle taste buds of European immigrants accustomed to the milder flavors of the European grape (*Vitis vinifera*). The human race has a long history of viticulture that goes back to the origins of civilization. Vine cultivation flourished in many regions of the world, and grapes were one of the most widely grown food crops. Ancient Egyptian mummies were entombed with the fruit as long as six thousand years ago, and grape remains have been found in Bronze Age dwellings in south-central Europe that are dated at 3500 to 1000 BCE.[2]

In the northeastern United States, the most widely cultivated grapes are derived from the common wild northeastern American fox grape (*V. labrusca*), which later got hybridized to make the more familiar 'Concord' grape. The other major parent for many of today's popular table grapes is *V. vinifera*, or the wine grape, which is native to southern Europe. Many generations of horticulturalists had failed to successfully grow wine grapes on the East Coast,[3] until the vines were hybridized to form a gene pool with multiple species and then combined with *V. labrusca*. Such hybrids are collectively referred to as *Vitis* × *labruscana* or the Labruscan grape.[4] Because *V. vinifera* hybrids sometimes lack cold hardiness and are susceptible to fungal diseases in North America, growers started grafting many of the plants onto *V. labrusca* rootstock, which is more resilient. If you have the space for a trellis or arbor, there is nothing like eating your own grapes right off the vine in late summer.

Growth Difficulty Rating: 3. Grapes can be challenging to grow due to their susceptibility to fungal diseases. The vines should be pruned every winter for the best grape harvests and to minimize disease problems.

Taste Profile and Uses: Grapes grow in clusters of fruit in variable sizes depending on the cultivar. The color of the outer skin as well as its interior flesh can range from a pale yellow-green to deep dark purple, burgundy, or almost black. Anyone who has ever eaten a ripe grape straight off the vine has tasted a range of flavors depending on the variety, from syrupy sweet and juicy to a sweet-tart taste. The fruits mature over a period of weeks, with individual clusters of grapes often ripening together at the same time. We love to eat the fruit standing right in front of the vine, but when there's an abundance of ripe grapes, it's best to harvest them or they can deteriorate fairly quickly on the vine. The fruits store well in a refrigerator, or whole bunches can be frozen. They are also great for making juices, jelly, and raisins. The ancient way to deal with an abundant crop is good old-fashioned winemaking, which also goes for many of the fruits in this book.

Plant Description: A woody, deciduous perennial vine that can grow anywhere from 6 to 15 feet (1.8–4.6 m) in a season. The trunk of the vine can be quite twisted and gnarled, with older specimens aging into a sculptural appearance that is adorned with flaky strips of peeling bark. The leaves on the *V. vinifera* cultivars tend to be prominently lobed with deep margins, while *V. labrusca* plants have less defined lobes along the edges of the leaves. The leaf edges on both species are lightly toothed, and foliage for both grows in a wide range of sizes up to 6 inches (15.2 cm) in length. All the vines produce coiling tendrils that wrap around the objects they grow across to help them climb toward light.

Flowers: Panicles of tiny white flowers grow in groups that are about 3 to 5 inches (7.6–12.7 cm) long and are produced on the current season's stem of the vine. The entire panicle of delicate flowers looks vaguely like a bottlebrush and is composed of tiny individual flowers that are roughly ⅛ inch (3 mm) long.

Pollination Requirements: Grapevines are self-fertile but will be more productive when another plant is sited nearby. The vines generally produce flowers in May when the temperature is between 59° and 68°F (15–20°C). The blossoms are pollinated by the wind and by insects.

Site and Soil Conditions: Grapes should be grown only in full sun; shaded locations will increase the potential for viruses flourishing on the fruit. The vines can tolerate a wide range of soil types, but they must be well drained. It is believed that the flavors concentrated inside the fruit reflect the minerals present in the deeper layers of the soil, hence the long tradition of wines being named for the specific geographic regions in which they grow. Providing deep but infrequent watering once a week the first year that the vines are planted will encourage the roots to grow farther into the soil to seek out water and nutrients, and the deeper the roots grow, the more drought-resistant the plant becomes—and the richer the flavor of the grapes.[5] Also, a greater amount of sun exposure on leaves toward the end of a fruit's ripening time will enhance the production of sugar in each cluster of fruit.

Hardiness: The hardiest grape cultivars are the hybrids of European vines that have been grafted onto native American rootstock. The vines are generally hardy to zone 5, or −20°F (−29°C), and various cultivars require a different number of frost-free days for fruit maturity. Several grape varieties can grow in zone 4 (−30°F/−34°C).

Fertilization and Growth: After the first year of being watered on a regular schedule, the vines will require less water as they age. If they receive a moderate amount of natural moisture from rain, they will actually benefit from being left alone. All grapes should be trellised on some type of support system for convenient harvesting, and be pruned once a season, because a wild rambling vine does not produce the best-quality fruit. The plants will also benefit from compost or leaf humus early in the spring as a mulch, but do not use

'Golden Muscat' flowers.

Young flower buds.

'Concord' fruit ripening.

'Golden Muscat'.

manure or any amendment with a concentration of nitrogen; this encourages vegetative growth at the expense of fruit production.

Winter Pruning Grapes: Grapevines bleed excessively and can be damaged when old growth (brown wood that is one year old or more) is cut, so major pruning is only done during the winter when the plants are dormant. Winter pruning increases the quality and size of the grapes, which is why all vineyards winter-prune their vines annually. Allow the plant to establish a central trunk with several main branches on each side of the trunk in its first growing year. After that the trunk and main branches are not pruned, and two to three thick buds are left on the ends of the main branches. This process is repeated every year, with more of the vine being pruned back to the original framework of the first year. This might seem extreme, until you understand that each bud you leave on the end of the main branch may produce up to 15 feet (4.6 m) of growth in a single season. During the part of the year that the vines are actively growing, you can remove unwanted new green growth from the current season as needed, but avoid cutting the branches that are one year old or older.

Cultivars: Because of the long history of grape growing, by 1909 there were over a thousand named grape cultivars.[6] Though the diversity of the varieties has diminished, a good selection of plants can still be mail-ordered through specialty nurseries, while some local garden centers tend to have the tried-and-true types.

Table Grapes (*V. labrusca*):

'Concord': Cultivated in the 1850s in Massachusetts and still one of the most popular grapes available. This is found in many grocery stores and farmers markets in late summer.

'Mars': A great cultivar for areas with hot and wet summers because it is very disease-resistant. The fruit also resists cracking and has a "foxy" flavor, similar to 'Concord'.

'Niagara': The green version of 'Concord', with a milder and slightly less astringent flavor, but it can be susceptible to fungal diseases.

Wine Grapes (*V. vinifera* × *V. labrusca*):

'Golden Muscat' (Hamburg × Diamond): A Cornell University introduction that we grow. It produces large, deliciously sweet golden-green grapes with a sharp, tangy finish.

'Himrod': A seedless yellow grape with good flavor developed in New York in 1952 that is hardy to zone 4 (−30°F/−34°C).

Propagation: Grapes are simple to propagate from hardwood cuttings. The best time to do this is in winter when you have to prune the vines, since you can use all of the discarded plant material to make new plants.

Pests and Problems: Besides birds and wildlife eating your crops, the biggest problem is the fruit's susceptibility to fungus and bacterial infections. Properly pruning your grapes will help prevent some of the problems, because thick, overgrown vines will shade fruit and foster mold growth. Strategically removing some of the leaves that shade the fruiting clusters can be helpful for creating better air circulation and drying any moisture on the developing fruit. Some of the worst grape problems are fungal, such as powdery mildew (*Uncinula necator*), downy mildew (*Plasmopara viticola*), gray mold (*Botrytis cinerea*), and black rot (*Guignardia bidwellii*). These are caused by environmental factors such as high humidity, wind, rain, and cool nighttime temperatures. Good cultural practices will help prevent diseases. These include: keeping vines and fruit off the ground so they tend to be dry; clearing all diseased and pruned plant material away from the vine (particularly shriveled, infected grapes); and weeding the soil around the vine's trunk so that the ground is less likely to harbor diseases.[7]

Brown Bag Covers: After several years of mediocre crops, we read about the use of brown paper bags as a form of protection. To do this, you wait one or two weeks after the vine has flowered and has formed tiny fruit. Then you take a small paper bag, such as a lunch bag, and slip it over a whole cluster of grapes. Double-fold the open portion of the bag together and secure the corners with staples such that the bag is attached to the vine with only the flower stalk inside it. The grapes develop and slowly grow inside the bag without being exposed to birds, insects, or most bacterial infections. The most important element in this technique is doing it right after pollination is completed because bacteria can damage the crops early on by attacking young, undeveloped fruit. The bags stay in place throughout the growing season. In late summer check for ripening fruit by making a small slit in the bag to inspect the grapes inside. The actual fruit does not require direct sunlight because it is the vines' leaves that use photosynthesis to produce mature fruit.

After many years of using organic spray fungicides that always failed, we have had several years of successful no-spray harvests with light losses using this simple method.

Hardy Orange

Poncirus trifoliata

Sour Orange, Trifoliate Orange

> *. . . used in the south for hedging because of its dense growth and thorny character; no sane person would attempt to penetrate this hedge! At one time I doubted its landscaping usefulness but have seen plants in Longwood's Mediterranean garden with yucca, prickly pear . . . that appeared to belong.*
>
> —Michael Dirr, 1998[1]

While it's fair to say that the great plantsman Michael Dirr does not love this plant, he thinks that hardy orange can be a useful and interesting landscaping addition. We, on the other hand, love *Poncirus trifoliata*, which is also known as hardy orange because its fruit pulp has a tangy, sour, citrus flavor. We especially love the sight of our contorted thorny citrus tree when it gets 16 inch (40.6 cm) of snow dropped on top of it in winter, as it takes on the bizarre appearance of a wild-looking snow cone. This demonstrates how tough this plant is and why this is the hardiest species of orange in the world. Although this is a crude citrus species that's mostly skin and seeds, it is still valuable as an edible ornamental with a wide range of culinary uses and can provide northern gardeners with another tough, pest-free edible plant for their home gardens.

Hardy orange is native to central and northern China and the Korean peninsula, where it has been cultivated since ancient times for medicinal uses and as a food flavoring. In the United States, hardy orange appears in plant literature as early as 1823, when it was sold at the Prince Nursery located in Flushing, New York, which was America's first nursery. William Saunders, a nineteenth-century botanist hired by the USDA, was the person responsible for popularizing the use of hardy orange in the South. Saunders provided hardy orange plants to growers throughout the southern states as a hardy rootstock for citrus.[2] Because some of the California orchards that grow citrus have been grafted onto hardy orange and exhibited viral problems in the past, it is no longer recommended as a rootstock in that state.[3]

Hardy orange became popular in the South as a tough plant for hedgerows. Although it's generally valued more as a protective hedge rather than for its fruits, we feel that hardy orange makes for both an interesting decorative addition as well as a useful edible plant in northern gardens.

Growth Difficulty Rating: 1. This citrus is surprisingly easy to grow and mostly pest-free. It grows in a variety of soil conditions.

Taste Profile and Uses: The fruit is a modified berry, 1½ to 2 inches (3.8–5.1 cm) in diameter, and its thick skin can be either glabrous (smooth) or slightly downy. In summer the small green fruits begin to develop and fill up the thorny branches. They turn yellowish orange in late September to October, and if hardy orange is allowed to ripen after picking for a few weeks, the fruit softens and the flavor is improved. The oranges are seedy and resinous, and although each fruit only contains a small amount of juice, they still have many culinary applications. We cut the citrus into thin slices to make a marmalade that was popular in nineteenth-century America.[4] The juice and pulp can also be used to brighten up bland fruit sauces; we especially like adding slices of the oranges to a jar of honey, which gives a distinct aromatic flavor to the sweetener. Both the rind and the juice are tasty additions to a cocktail, such as a whiskey sour or a margarita. In traditional Chinese medical practice, hardy orange is used for anti-inflammatory applications.[5]

Fall color and ripe fruits.

'Flying Dragon' in winter.

The attractive bark.

Flower buds about to open.

Flowers and thorns.

Plant Description: In the wild, trees can grow 6 to 20 feet (1.8–6.1 m) and form into a dense tangle of branches, but under cultivation most trees reach only 5 or 6 feet (1.5–1.8 m) in height. An ornamental plant with deep green branches that are curved and bent, and armed with 1- to 1½-inch (2.5–3.8 cm) wicked green thorns. Hardy orange has ovate-shaped foliage that is 1 to 2½ inches (2.5–6.4 cm) long; the leaves are arranged in groups of three, hence its taxonomical name *trifoliate*. These trees are the only member of the citrus family that are deciduous, which can provide some interesting fall color when the leaves develop touches of scarlet and yellow before falling off the tree. Because hardy orange has contorted, thorny branches, these plants provide interesting ornamental touches against snow, but it is best not to plant them near walking areas where people might brush up against the sharp curved thorns.

Flowers: The slightly fragrant blossoms appear in late April to early May and are produced on the previous year's branches. The 1½ - to 2-inch (3.8–5.1 cm) white flowers have five petals surrounding a yellow stigma and anthers.

Pollination Requirements: Hardy orange is fully self-fertile and productive, and mature specimens of the trees are often completely covered in fruit in the fall.

Site and Soil Conditions: Like all citrus these plants prefer a full-sun exposure to produce ripe fruit. Our tree has performed well in average soil with a heavy clay content; indeed, some sources say that hardy orange does best in heavy clay soils as long the planting site does not contain a heavy amount of salt and isn't permanently wet.[6]

Hardiness: In plant literature the cold tolerance of hardy orange seems confusing, with some sources listing it as hardy to zone 7, or 0°F (−18°C), but other sources say the tree can tolerate temperatures down to zone 5, or −20°F (−29°C).[7] Our hardy orange cultivar 'Flying Dragon' has proven tough and resilient, surviving many long, brutal winters with several days of temperatures down to −8°F (−22°C) with little damage. We do think that small-sized trees should be protected before being planted in the colder regions; once the bark gets thicker and strong root systems develop after a few years, they should be fairly resilient.

Fertilization and Growth: Although hardy orange seems tough and grows without any help as long as it is in average-quality soil, we have found a yearly application of an organic citrus fertilizer will help the plants be more productive. In our gardens, citrus is the one plant family that receives a commercial acidic plant fertilizer, which we give prior to the foliage leafing out, and once again in midsummer.

Cultivars: There is a contorted cultivar of the plant and a few hybrids of hardy orange that are crossed with other species of citrus that are called *citrange*, although these seem difficult to find in nursery cultivation.

'Benton': A citrange hybrid of hardy orange crossed with *Citrus sinensis* that originated in Australia, reportedly hardy to zone 7, or 0°F (−18°C).

'Flying Dragon': An attractive form of hardy orange with a contorted growth habit, curved spines, and bowed branches, that grows a little shorter than the straight species.

'Morton': Another citrange hybrid form of hardy orange crossed with *C. sinensis*. This cultivar is one of the hybrid fruits created by Walter T. Swingle in the 1890s.[8] The tree produces larger fruit with fewer seeds than the *P. trifoliata* species and is rated hardy to zone 8, or 10°F (−12°C).

Related Species: Hardy orange has always been viewed as monotypic, or the only species in its genus. A new species of this genus was discovered in the Yunnan province of China in the 1980s. This plant is now called *P. polyandra* and is considered to be an endangered species.[9]

Propagation: These are easily grown from seeds, with a high germination rate if they are planted while fresh. Late-fall cuttings of hardy orange are sometimes successful, and the rate of success is increased if root hormone is applied to the bottom of the cuttings. We have found that the cuttings are slow to root and may take a few seasons to exhibit any noticeable new leaf shoots. Air-layering with a small bag of soil tied around a stem has also worked to produce rooted plants.

Pests and Problems: We have not seen any major pest or problems with our tree. Our plant has not produced any volunteer seedling plants, but in the southern parts of the United States, hardy orange has become an invasive species.[10]

Hazelnut

Corylus americana, C. avellana

American Hazelnut, European Hazelnut, Filbert

> *Hazels belong to a very ancient family, and some of the fossil hazel leaves are hardly to be distinguished from those species which are living today.*
>
> —Robert T. Morris, 1921[1]

The Latin name *Corylus* is derived from the Greek word *korys*, which means "helmet" or "hood," in reference to the fleshy husk that completely covers each nut.[2] There is a long history of people using hazelnuts as a nutritious food crop in addition to the nut being an important forage for wildlife. The name *cobnuts* refers to larger cultivated varieties of hazels, and *filberts* is often used as the name for European hazelnuts grown in the Pacific Northwest. But in truth, the word *filbert* was a corruption of *full beard*, which specifically refers to a hazel whose husk extended past the nut.[3]

European hazelnut (*C. avellana*) is indigenous to Europe and the United Kingdom, and is the main source for the majority of improved cultivars. Although there are tree forms of the hazelnut, the common form is a dense, multi-stemmed shrub that sends up new shoots every year. Because of its typical form of growth, there is a long tradition of hazelnuts being planted as a multifunctional hedgerow that is not only used as a food source but also is regularly pruned to create a sustainable supply of wood for fires and basket making, as well as providing a protective windbreak around farm fields. The dominant species of hazelnut native to the United States is the American hazelnut (*C. americana*).

Growth Difficulty Rating: 1. Both American and European hazelnuts are easy to grow, but the latter can be susceptible to EFB, or eastern filbert blight (see the "Pests and Problems" section below).

Taste Profile and Uses: The nuts are covered by an involucre (protective husk), which is fleshy as well as decorative. As the nuts ripen through the summer, this involucre expands until the nut is ripe and falls from the husk, which sometimes remains on the tree branch. The nut size varies from ½ inch (1.3 cm) in most wild species to 1 inch (2.5 cm) in bred cultivars. The thickness of the shells varies with each species and cultivar, but they're generally easy to open with a nut cracker.

Hazelnuts are a nutritious nut, with a mild sweet-buttery flavor that we think tastes delicious eaten out of hand, but roasting the nuts brings out their best culinary qualities and they are often used this way as an addition to both savory and sweet recipes. One of the most popular products made from this nut is Nutella, the famous chocolate hazelnut spread.

Plant Description: A moderately fast-growing, deciduous, multi-suckering shrub or tree that can reach 8 feet (2.4 m) (American) to 20 feet (6.1 m) (European) in height, depending on the species and cultivar. Both species have bright green foliage that has an attractive pleated appearance with toothed edges. The leaves are alternately arranged on the stem. The 2- to 4-inch (5.1–10.2 cm) European leaves are more ovate, while the American are 2½ to 6 inches (6.4–15.2 cm) long, with a narrower shape and pointed tip. As the plants get older, the root sections send out new suckers in a circle surrounding the bush, sometimes appearing a little untidy and messy.

Flowers: Hazelnuts have both male and female flowers on the same plant, with very attractive catkins (male flowers) that bloom to a length of 2 to 4 inches (5.1–10.2 cm) in the American species and 2 to 5 inches (5.1–12.7 cm) for the European types. Both the American and European hazelnuts produce yellow catkins, while

The male flowers of 'Rote Zeller'.

The tiny female flowers on an American hazelnut.

'Rote Zeller' leaves transitioning to green.

'Geneva' involucre-covered nuts.

purple-leaved cultivars produce light pink catkins. The catkins for both types develop in the late summer and are displayed all winter long on the deciduous branches, creating a highly ornamental effect, with the long flowers opening up into handsome tassels in early spring. Shortly after this the tiny scarlet-red female flowers for both species open near the end of the branch stems and grow to about ⅛ inch (3 mm).

Pollination Requirements: The flowers open in early spring when few insects are alive, so hazelnuts depend on wind to be pollinated. Bushes are not self-fertile and must have another hazel bush to cross-pollinate them to produce nuts. They benefit from being planted in groups, and American and European species will cross-pollinate each other, which can provide an opportunity to try growing different types of hazels. There are several new cultivars that open at slightly different times and may have specific pollination requirements.

Site and Soil Conditions: Both plant species are widely adaptable to different growing conditions but are best grown in slightly acidic soil in part shade, or in consistently moist soil in full sun. Hazel bushes do less well in sandy soils, but adding leaf humus, compost, or other organic materials will help the plant's roots adapt. American hazelnut is found in our region of upstate New York in forest clearings and along the alluvial clay soil banks of streams, as long as they are not waterlogged, whereas European types prefer organically rich, well-drained soil.

Hardiness: American and European plants are hardy to about zone 4, or −30°F (−34°C), but different cultivars have various levels of cold tolerance.

Fertilization and Growth: Underlining the plant's toughness, hazel bushes produce edible nuts in the wild in bad soil with little fertility. But plants will be more productive with regular additions of compost applied as a mulch.

Cultivars: Since the early 1900s, efforts to develop suitable hazelnut cultivars for eastern North America have been ongoing, with the goal to combine the disease resistance and winter hardiness from cold-climate-adapted species with the larger kernels and the thinner shells of European hazels.[4] The two primary sources for hazelnut breeding have been American and European hazelnuts, which have been combined and crossed with other hazel species to produce close to one hundred cultivars.[5] Below we list a few of the most recommended cultivars that we have found produce reliable crops. In the last fifty years, the big push in breeding has been to increase nut yields and build immunity from EFB in order to create disease-resistant nuts. So in the near future, there should be many new blight-resistant hazelnuts on the market.[6] It should be noted that because of the danger of EFB, mail-order nurseries will not ship hazelnuts to certain parts of the United States.

American Hazelnut:

'Winkler': An American cultivar introduced in 1918 that grows to 6 feet (1.8 m) and is not susceptible to EFB, with nuts that are slightly larger than straight species.

European Hazelnut:

'Rote Zeller': An ornamental purple-leafed cultivar from Germany that is hardy to zone 5 (−20°F/−29°C). The plants begin each season with scarlet-burgundy leaves that turn green by the middle of summer. They yield medium-sized nuts covered with attractive scarlet husks.

Hybrids:

'McDonald': A large bush that has good EFB resistance and was introduced by the plant breeder Peter McDonald. It has good-quality nuts and is hardy to zone 4 (−30°F/−34°C).

'Slate': A blight-resistant (possibly immune), productive cultivar with very large and flavorful nuts. Introduced by George Slate of the New York Agricultural Experiment Station in Geneva, New York, the plants reach 12 to 18 feet (3.7–5.5 m) in height.

Related Species: The beaked or Western hazelnut (*C. cornuta*) is a bush native to large portions of North America from Maine to British Columbia. A drought-resistant, 4- to 9-foot (1.2–3 m) tall bush, though it can grow larger.

Propagation: Nuts can be gathered in late summer (before squirrels get them) and planted in a pot over the winter in a protected site that rodents cannot find. A traditional way to get plants that are true to the original cultivars is by digging up and replanting some of the new shoot suckers that form around the base of the bush, which is best done in early spring. Another way to propagate hazels is by layering branches: Bury low-lying branches in the soil to allow them to form roots. Roots may take more than one season to develop, and you can leave buried limbs in place for years before digging them up so that you get plants with substantial root systems.

Special Note: Grafting is another method of producing cultivars. The chief problem with a grafted hazelnut is its propensity for suckering, and if these suckers are not pruned away from the tree, the resulting shoots will not be true to the specific cultivar. For this reason we recommend planting only bushes that have their own roots and are not grafted.

Special Pruning Considerations: As hazelnut branches age they will eventually stop yielding nuts. This means that older, less productive branches should be pruned off the bush in favor of younger ones that will replace them. Since nut productivity on branches lasts from five to ten years, look for branches that have stopped producing catkins to determine which should be pruned off. These are coppiced (cut back to the ground periodically to stimulate new growth). Many specimens of hazel have lived for hundreds of years in Europe.[7]

Pests and Problems: The chief limiting factor to growing hazelnuts in the United States has been eastern filbert blight (EFB). This disease is a naturally occurring fungus, native to the eastern portion of the United States and a few sites in the West. Many university and USDA breeding programs have been dedicated to producing disease-resistant cultivars.[8]

EFB is the major problem with European hazelnuts. The blight (*Anisogramma anomala*) has evolved alongside American hazelnuts and does not affect this species. European hazelnuts were introduced into the United States as a commercial crop in the early twentieth century, and shortly thereafter entire East Coast orchards were destroyed. EFB was found in various states east of the Rocky Mountains, and in the 1970s it was discovered in orchards in Oregon.

EFB appears as oval cankers that grow in rows on branches; by summer, girdled branches become obvious because dead leaves remain attached. Over a short span of time, more leafless, dead, and dying branches become visible in the canopy as the disease spreads. Diligently pruning out infected branches and applying a fungicide will help to reduce the disease in an affected area. Dispose of diseased branches by burning them or placing them in the garbage.

If EFB is a major problem in your region, selecting disease-resistant cultivars will be crucial to reducing the risk of the blight. In our gardens the ten hazelnut bushes of different species and disease-resistance levels have been growing for over eighteen years and we have not had any problems with the blight. However, we do live in a part of the United States where EFB is present, and we are always on the lookout for signs of the disease.

Heartnut

Juglans ailantifolia var. *cordiformis*

Japanese Walnut

> *While I do not suppose that these Oriental walnuts will ever become of any considerable commercial value, they are worth planting for shade and ornamental trees. They are rather precocious, coming into bearing at an early age, and the nuts are not only edible, but will always be an acceptable addition to the unimportant although agreeable household supplies.*
>
> —Andrew Samuel Fuller, 1896[1]

The Japanese walnut (*Juglans ailantifolia*) was first introduced in cultivation under the Latin name *J. sieboldiana* and is sometimes referred to as the seibold or cordate walnut. The tree was imported into the United States in 1870 as a fast-growing, exotic shade tree with ornamental compound leaves that just happened to produce hanging clusters of edible nuts.[2] The trees were widely distributed throughout the United States and Canada and proved to be very adaptable, growing in areas that experienced both hot and cold temperature extremes.[3]

The Japanese walnut produces nuts with two distinctive shapes: those that are oval-round, and those with a unique heart-shaped outline. This confused late-nineteenth- and early-twentieth-century botanists, who incorrectly believed that the trees that produced the heart-shaped nuts were a different species, and named it the heartnut or *J. cordiformis*, which is Latin for "heart-shaped."[4] Since the Japanese walnut was new to North America, there was a lot of confusion about the correct scientific name for this species until the trees were correctly identified by the great plant explorer and head of the Arnold Arboretum, E. H. Wilson, who confirmed that the seibold walnut was the wild Japanese walnut and that *cordiformis* was a cultivated form of the same species. This unusual form of the tree was probably the result of a sport (genetic mutation) that bore distinctive flattened, heart-shaped nuts.[5]

In the early 1900s horticulturists like J. F. Jones championed the planting of heartnuts and propagated a number of improved varieties, to encourage citizens to plant the trees. Jones felt that heartnuts featured many great agricultural attributes because the trees grew fast and tended to yield crops a few years after planting, and were less affected by pests. Plus the nuts lacked the bitter tang that many walnuts can have.[6] Growers began planting Japanese walnuts and sold young seedling trees, but many of the seedlings were accidentally hybridized nuts that were the result of cross-pollination with the Butternut (*J. cinerea*), which flowered at the same time as the heartnut. These two species crossed freely, so many of the heartnut seedlings ended up being white walnut hybrids, and most horticulturists stopped promoting the trees. Heartnut declined in popularity.[7]

By the 1940s growers still knew little about this tree, and at one point the Arnold Arboretum's bulletin requested information from any members of the public who were planting improved heartnut cultivars,

Heartnut foliage.

so they could keep records of the trees' adaptability to cold northern climates and pass on that information to their local agricultural experiment stations and the Northern Nut Growers Association.[8] Now almost 150 years after being introduced into the United States as a tree with a lot of potential, the heartnut is still not widely appreciated, despite all the wonderful things it has to offer as a delicious and flavorful food crop.

Growth Difficulty Rating: 1. Heartnut trees are easy to grow, with only minor pest problems.

Taste Profile and Uses: The oval or heart-shaped nuts are covered in a green husk and hang on long racemes in bunches of two to twenty, which mature in late summer. Under the shuck (husk), the smooth, light brown outer shell is ovate and slightly flattened with one pointed tip. When cracking open the cultivated variety, 1- to 1½-inch (2.5–3.8 cm) shells, it looks like a flattened heart-shape outline, with a slight ridge dividing the shells front and back. The nuts are said to be thin-shelled compared with the thicker-shelled black walnut (see "Black Walnut" chapter) and are easier to crack. The heart-shaped kernels are plump and light gold in color, although nuts that have a uniform pale gray or beige tone are also fine to eat. The flavor has a mild nutty taste with no bitterness and is often compared to a butternut. The flavor of the shelled nuts will actually improve in storage at room temperature and peaks between five and eight years after harvest. Beyond ten years at room temperature, heartnuts may become slightly stale, but they are still good enough to use.[9]

Plant Description: Heartnut is a deciduous tree growing up to 60 feet (18.3 m) tall, but more commonly grows to 30 feet (9.1 m), and the trees tend to grow as wide as they are high. The leaves, which are composed of eleven to seventeen pinnately compound leaflets arranged along the opposite sides of the stem, can reach up to 3 feet (91 cm) in length, giving the trees a tropical appearance. Each leaflet is light green and slightly hairy, and grows between 2 and 6 inches (5.1–15.2 cm) long and up to 2 inches (5.1 cm) wide.

The trees also have a good reputation for yielding crops at a younger age than many other members of the walnut family. Seedling trees often bear crops when they are five to seven years of age, with grafted trees often yielding crops in one to three years.[10] In our gardens the two heartnut trees began to flower after five years and produced immature nuts after seven years.

Flowers: The heartnut is monoecious producing both male and female flowers on the same tree. The male flowers are the yellow-green catkins, which typically grow 6 to 10 inches (15.24–25.4 cm) long. The female flowers are produced with the flush of the current season's leaf growth and are shaped like small pea-sized ovals, topped with scarlet red pistils for receiving the pollen. Pollination is carried out by the wind. The trees produce large quantities of flowers in spring at the same time that new leaves appear.

Pollination Requirements: Heartnut trees are only partially self-fertile and a single tree can bear small crops, but two or more trees planted together will produce a lot more of the nuts. The male and female flowers on individual trees open at different times and so planting two trees together will help cross-pollination. Straight seedling heartnut trees are compatible to plant together for good pollination, but if you plan to use grafted heartnut cultivars, it is recommended that you match an early-flowering type (early pollinator) with a late-flowering type (late pollinator) in order to correctly time the flower cycles.[11] Specialty nurseries that sell grafted heartnuts generally list which varieties are compatible, and there is a list of suitable pollinators in the "Cultivars" section below.

Site and Soil Conditions: Heartnuts prefer moist, fertile, well-drained soil, though they will tolerate a wide range of conditions, including areas with sandy soils as well as heavy clay content. Heartnuts require a full-sun site for the best nut production, and the trees cannot tolerate growing in waterlogged areas.

Hardiness: The trees are rated hardy to zone 5 (−20°/−29°C), but the flowers open in early spring and may occasionally be damaged by late frosts in growing areas colder than zone 6 (−10°/−23°C), since the trees are terminal-bearing (flowers in the buds on the ends of branches), and these are the parts of trees most vulnerable to extreme temperature fluctuations.[12] But there are records of heartnut trees surviving temperatures of −30° to −40°F (−34° to −40°C), zones 3 and 4.[13]

Fertilization and Growth: Heartnut trees are heavy feeders and should be given compost or leaf humus in early spring as the trees come out of dormancy, and fertilized once again in late summer before the nuts fall. The husk and the shell reach their full size early in the year, and for the rest of the growing season the kernel develops inside the shell from the carbohydrates generated by the leaves. So if there is heavy foliar damage by insects, drought, late frost, or an early freeze, any of these problems (or a combination of them) can affect whether or not the nuts develop inside the shells.

The tiny female flowers.

The male flowers.

The handsome bark on a juvenile tree.

The heart-shaped kernel.

Cultivars: It seems that in almost every decade, new varieties are developed by nut breeders who continue to select cultivars for thinner-shelled nuts with freely releasing kernels. There are dozens of different Japanese heartnut cultivars; although these trees are rare in the nursery trade, a few specialty mail-order catalogs sell grafted trees.[14] If you are planting a grafted variety, it is important to match an early-flowering with a late-flowering variety.

'Campbell CW 1': A late pollinator that is very hardy with easy cracking quality and good-quality nuts.

'Campbell CW 3': An early pollinator, a heavy producer, and one of the hardiest heartnuts.

'Imshu': A late pollinator yielding mild-tasting large nut kernels, with exceptional cracking quality.

'Locket': An early pollinator that produces large-sized nuts, with freely dropping whole kernels.

Related Species: Heartnut relatives include the butternut (*J. cinerea*), as well as the buartnut (*J.* × *bixbyi*), which is the heartnut crossed with a butternut.[15]

Propagation: Seedlings are the most common way to propagate heartnuts, but random seedlings often grow into trees whose nuts do not resemble their parents because heartnut is a seedling cultivated from wild trees. Though these nuts can be delicious and flavorful, many heartnut seedlings revert to wild Japanese walnut trees, producing nuts that are thick-shelled and harder to crack than the cultivated type. Most seed-grown trees produce harvests where 70 percent of the nut seedlings revert into wild-type trees, but there are heartnut trees that produce heart-shaped nuts. This tends to be rare.

Heartnuts are among the most challenging trees to graft, requiring great skill for success, but some of the better heartnut cultivars can be grafted onto black walnut rootstock.

Pests and Problems: These trees are usually fairly pest-resistant, but all the members of the walnut family can be attacked by several insects that can affect the nuts, foliage, and branches. Our trees have had small wounds on some of the young stems from insects chewing on them, though these appear to be minor cosmetic problems and do not seem to be causing great harm to the trees.

One of the biggest problems in growing heartnuts is that the kernels can fail to fill up, and the two most common reasons for this are a short growing season and an insufficient amount of summer heat to ripen the nuts.

Highbush Blueberry

Vaccinium corymbosum

> *Blueberry breeding has now been carried on for 10 years, with the result that instead of berries the size of peas, like the ordinary wild blueberry, we now have hybrids producing berries the size of Concord grapes.*
>
> —Frederick V. Coville, 1921[1]

Blueberries and their many relatives across the world have been gathered from the wild by ancient Indigenous cultures in Asia, Europe, and North America for thousands of years, which is exemplified by the discovery of a preserved, two-hundred-year-old Inuit sealskin pouch filled with blueberries, that was uncovered in Canada.[2] Wild blueberries grow in many parts of the United States and remained a wild-harvested food that began to grow in popularity through the 1940s.[3] But blueberries were turned into a major agricultural product due to the work of two people: Frederick V. Coville and Elizabeth White. After many horticulturalists failed to successfully cultivate blueberries, Coville started conducting breeding experiments on the fruit in 1909 on behalf of the US Department of Agriculture. His work on breeding wild blueberries led to the major discoveries that allowed them to be successfully cultivated, specifically the necessity for growing plants in acidic conditions, and the discovery that blueberry plants have difficulty collecting nutrients out of the soil and require a beneficial mycorrhizal fungus surrounding their root systems to break down nutrients for them.[4]

Coville's experiments were conducted in a pine barrens region in Whitesbog, New Jersey, through a partnership with Elizabeth White, who provided a portion of her family's farmland that had acidic soil.[5] From 1906 to 1937 they conducted experiments in hybridization and selected the best blueberry varieties with a focus on the improved size, flavor, color, ease of picking, and keeping qualities, as well as an emphasis on lengthening the blueberry picking season by producing early-, midseason-, and late-maturing fruit.[6] These experiments yielded the first improved cultivars with large fruit, which were then hybridized by other berry breeders into the plants that now grow on modern blueberry farms and in backyard gardens.

Growth Difficulty Rating: 1. Blueberries are easy to grow with consistent moisture and acidic soil or amendments to acidify the soil around the plant's root system.

Taste Profile and Uses: The highbush blueberry (*Vaccinium corymbosum*) produces a false berry, which is a fruit with an inferior ovary that includes other parts of the flower such as the basal parts of the sepals, petals, and stamens.[7] It is about ¼ to ½ inch (0.6–1.3 cm) in size, but the fruit on bred varieties can grow larger, with a few cultivars that produce berries almost the size of a quarter. The fruit on each bush does not ripen all at once but matures sporadically over a period of time, ranging from a few weeks to more than a month. The fruit has a rich, distinctive flavor and despite the numerous cultivars available most highbush blueberries have a similar taste, but with some variation in the amount of sweetness in each berry.

Blueberries are good for dozens of uses such as fruit preserves, cereal, yogurt, pies, muffins, ice cream, and in salads, but are also one of the best berries for fresh picking right off the bush.

Types of Blueberries: There are two other major species besides highbush that are sold as a commercial crop and used in plant breeding—lowbush (*V. angustifolium*) and rabbiteye (*V. ashei*)—and two hybrids, southern highbush and half-high blueberry. This chapter is focused on highbush blueberry, but all of the plants listed above have been cross-bred with highbush to generate varieties that produce different-sized fruit, along with different climatic preferences, bush heights, and fruit that ripens at different times of the season. Highbush blueberry grows wild along the East Coast and from Maine across to Minnesota, then down to North Carolina, although some sources say that highbush can be found as far south as Florida and Louisiana.[8]

Plant Description: Most nurseries sell highbush blueberry plants that are the bred varieties and grow 4 to 6 feet (1.2–1.8 m) high and wide, but in the wild, blueberry has a great variability in its growth habit that can range from 4 to 12 feet (1.2–3.7 m) high. The bushes grow slowly with large, fibrous root systems that spread out and produce more branches, which ultimately form into multi-stemmed shrubs with deep green leaves that are lance-shaped.

Blueberries can provide decorative flourishes to any garden for all four seasons. In spring the flowers are delicate and attractive, then in summer the berries change from light green to an array of blue tints ranging from powdery light blue to a dark navy hue, with some cultivars turning bright pink. In autumn the leaves turn a spectacular range of reds and purples as they develop fall color, and throughout the winter young branches remain an attractive red to red-brown color.

Flowers: Small bell-shaped flowers have five fused petals that hang down. The flowers are white or white with a pink tinge and are about ⅓ inch (9 mm) long. Some new cultivars have pink blossoms.

Pollination Requirements: Blueberry bushes are pollinated by bees in early spring and always produce better crops when planted in groups. Though many blueberry cultivars are listed as being self-fertile, the yield will always be small on bushes planted alone, so the best way to grow bountiful harvests is to plant several different varieties.

Site and Soil Conditions: Blueberries thrive in full sunlight when their root systems get consistent moisture. They are one of the few fruits that prefer to grow in acidic soil with a pH value of 4.5 to 5.5. The plants' root systems will tolerate both sandy and clay conditions as long as they are planted in well-drained soil that is enriched with acidic organic amendments such as decomposed oak leaves, or untreated pine mulch or bark.[9] But our favorite acidic amendment is the "duff" (plant litter) beneath evergreen trees, which will not only acidify the soil but is more likely to contain living beneficial mycorrhizal fungi as part of the soil's organic elements. We also use pine needles as a thick mulch to suppress

The colorful flower buds.

Highbush blueberries.

Fall color.

The winter branches.

weeds, and the needles provide nutrients to the soil as they break down over time. Blueberries do not produce bountiful harvests if their root systems are constantly dry, so it is important to provide the plants with some supplemental water if you are in an area that does not get rainfall throughout the growing season. For this reason, it is very important to keep young bushes well watered the first couple of years after planting; after this the plants can tolerate longer periods of drought between rainstorms.

Hardiness: Wild highbush blueberry plants are very hardy and are rated as zone 3, or to −40°F (−40°C).[10] Because named varieties are often crossbred out of multiple species, they are slightly less hardy and listed as zone 4, or to −30°F (−34°C). But when purchasing plants, make sure to check the cold tolerance for individual cultivars.

Fertilization and Growth: Blueberries thrive in acidic conditions, so top-dressing the soil around the roots with pine needles, pine sawdust, or pine bark should prevent them from drying out completely. Such amendments will decompose over time and give the soil more acidity, creating a better growing environment for the blueberry's roots. You can also use a commercially produced organic amendment sold in nurseries for fertilizing rhododendrons and other acid loving plants. On large, established shrubs the branches that are nine years old stop producing fruit, so these should be pruned off a few at a time each spring to not only encourage air circulation but also allow new branches to replace them.[11]

Cultivars and Seasonal Ripening Periods: Highbush blueberry varieties have been selected for different ripening periods, and the broad spectrum of all the cultivars can extend fruit harvesting for more than a three-month period. These ripening times will vary for every gardener depending on a series of environmental factors—geographic location, hardiness zone, seasonal temperatures, and amount of sun exposure. The following are some of the best-known varieties of highbush blueberry, grouped by their fruit ripening times in the order that the fruit matures, but all blueberry bushes produce fruit over a period of several weeks, so the list reflects the most common period in which the ripe fruit can be harvested. Nurseries, farmers, and plant breeders classify highbush blueberries into three main groups that reflect when the fruit can be picked off bushes. The best chance for successfully harvesting blueberries over a long period of time is to choose at least three to four (or more) varieties and try to select types in a few of the separate seasonal ripening periods listed below.

Early Season, June through the end of July: 'Earliblue', 'Bluetta', 'Duke', 'Spartan', 'Patriot', 'Northland', 'Collins', 'Blue Jay'.
Midseason, July through mid-August: 'Blueray', 'Draper', 'Toro', 'Hardyblue', 'Bluegold', 'Chippewa', 'Bluecrop', 'Chandler'.
Late Season, Late July through early September: 'Legacy', 'Jersey', 'Coville', 'Darrow', 'Lateblue', 'Liberty', 'Elliot', 'Aurora'.

Related Species: *Vaccinium* is a giant genus of plants with over 450 species distributed all over the world.[12] While we grow over a dozen different cultivars and love the berries of highbush, we still believe that the smaller-sized American lowbush blueberry (*V. angustifolium*) has a stronger and more intense taste and a truer "blueberry" flavor. It is unfortunate that most people do not encounter lowbush blueberry fruit except as the packaged, frozen "wild" blueberries in supermarket freezer aisles.

Propagation: Cuttings taken during the winter and grown out for a year in a shaded environment will produce new plants.[13] But an easier way to propagate is to dig up firm branches in early spring from the base of an established bush with some of its roots still attached, and plant them in pots with acidic potting soil for two or three seasons.

Pests and Problems: The universal pests for blueberries are birds and chipmunks, which relish the fruit so much that they can strip bushes clean of both ripe and unripened berries in a few hours. The best remedy for this assault is to cover bushes with netting around the plant that can be lifted or rolled back for easy access to the fruit.[14] Each year we solve this problem by erecting a temporary blueberry cage—a large square frame covered over with bird netting—which we remove after the fruit is harvested.

Blueberries are generally disease-free but can be affected by cankers from a fungus (*Godronia cassandrae*) and blueberry stem galls (*Hemadas nubilipennis*), both of which can be pruned off the plants.[15] Blueberry bushes are also more likely to develop root-related diseases if grown in waterlogged soil.[16]

Himalayan Chocolate Berry

Leycesteria formosa

Flowering Nutmeg, Himalayan Honeysuckle, Pheasant-Eye

> *Native to the Himalaya in shady forests; introduced in 1824. This handsome shrub likes a rich soil, and in spite of its natural habits, a sunny spot. Birds, especially pheasants, are very fond of the berries, for which reason it is sometimes planted as covert.*
>
> —W. J. BEAN, 1914[1]

The Himalayan chocolate berry is relatively unknown to growers in the United States. We have only encountered it in a few botanical gardens, such as the Untermyer Gardens in Yonkers, New York, where it was planted as an ornamental annual. Not only are the plants seldom seen in gardens, but the striking fruit from the shrub is almost never commented upon in garden literature.

This plant is endemic to the forest areas of the Himalayas, western China, and eastern Tibet, and because it was widely used in Europe and Australia as a shrub and border plant, it unfortunately escaped from cultivation and has invaded forests in regions with warm, moderate climates.[2] The shrub was originally introduced into cultivation in Britain and Ireland in 1824 as an ornamental bush. Landowners were delighted by the decorative attributes of a hardy plant that had the ability to rapidly naturalize large areas; the plants were also used in country estates as foraging shelter for game birds.[3]

'Golden Lanterns' chocolate berry.

The Himalayan chocolate berry is a member of the honeysuckle family (Caprifoliaceae) that became popular with urban gardeners in the 1850s as a Victorian shrubbery plant because it combined a dense growth habit with attractive dangling flowers that offered a long period of seasonal beauty. We began seeing this plant offered in the United States about ten years ago in mail-order nursery catalogs as a showy shrub with edible fruit.

The harsh winter temperatures of the northern parts of America have kept the plant's invasive tendencies in check, so gardeners can enjoy the beautiful flowers and distinctive edible berries without worrying about it being an invasive species. While the plants do not produce a substantial crop of berries, they do offer small nibbles to the gardener who walks by them, and what a taste! Though the fruit is small and messy, we have never had a berry that has a flavor profile quite like chocolate berry, which tastes like dark chocolate, espresso, burned caramel, and blackberry—all at the same time. So, although these plants are not going to provide a major harvest for food canning, they are worth growing for gardeners and homeowners with an adventurous palate who want a diverse garden with many flavors.

Growth Difficulty Rating: 2. These shrubs are relatively easy to grow, but we recommend giving some protection in areas where the winter temperature dip below 15°F (−9°C).

Taste Profile and Uses: The ¼-inch (6 mm) fleshy berries grow in clusters, which do not ripen as a single crop but mature sporadically across the whole bush. The young berries start off green and develop color as they mature, turning maroon until they are almost black when ripe. By early fall the ripe, dark berries have a soft, delicate texture and they frequently split open when you pick them. The taste of the berries is very complex and combines many unusual flavors such as bittersweet chocolate, burned caramel, and blackberry, although others who have tried the fruit get hints of coffee and red wine.

There are several reasons why the chocolate berry may not be well known as a fruit. Primarily it's because the tiny berries ripen unevenly and are too perishable for the marketplace due to their tendency to crack open when fully ripe. So you never have more than a handful of fruit—not enough to make it a worthwhile commercial crop.

Plant Description: Chocolate berry is a multi-caned deciduous subshrub in the honeysuckle family. It is a fairly vigorous plant and in the proper conditions will grow up to 6 feet (1.8 m) tall in a season. The shrub's dark green leaves are ovate with lightly scalloped edges that end in a long-tapered point. The leaves can be anywhere from 3 to 8 inches (7.6–20.3 cm) long and grow on opposite sides of the plant's hollow stems. The overall look of the plant is very decorative with attractive leafy, arching branches that form a large, fountain-like form with an abundant display of flowers and fruits.

The plant is considered a sub-shrub because the roots are cold-hardy and regrow new shoots each season, but the branches will live or die back each season depending on the amount of cold exposure. Thus, in warm garden areas the plants are considered shrubs with branches that live all year, but in colder regions the branches die back to the ground and the plants behave like a perennial.

Flowers: The flowers on this shrub are beautiful and are produced in large quantities that start in early spring and continue right through fall. The 2½- to 3½-inch (6.4–8.9 cm) blooms are bell-shaped with pink-tinged edges and bloom on new growth. The pendulous clusters of flowers do not have pedicles (a small stem), and instead droop from the axils of the leaves. Each flower is surrounded by large bracts that start off green with burgundy-red veining and progressively turn into a darker burgundy as the berries start to form. The deep burgundy bracts are a striking aspect of this species, and shrubs display flowers at different stages of development. Even each individual fruiting cluster has fruit that is developing at different stages of maturity, ranging from green to black.

Pollination Requirements: The chocolate berry is self-fertile, with the pollination carried out by bees, wasps, and other insects swarming over the pollen-rich flowers.

Site and Soil Conditions: Although found in the woodlands of its native Asian lands, this plant flowers more prolifically when planted in full sun but can tolerate some shade. The shrub can be grown in sandy loam or areas with a good amount of clay content, but it prefers rich soil that is well drained with consistent moisture throughout the growing season. If the planting area experiences long periods of drought, the chocolate berry may require some supplemental watering.

Hardiness: The cold tolerance of chocolate berry is rated at zone 7, or to 0°F (−18°C), but the root systems can often survive in zone 6 (−10°F/−23°C) with winter protection. In parts of the country where the temperature goes below 15°F (−9°C), the branches die back to the root systems each winter. Before winter we strongly

Flowers amid foliage.

The showy bracts around the flowers.

Ripe and unripe berries.

Colorful clusters of berries.

recommend burying the first 6 inches (15.2 cm) of the base of the plant with a heavy mulch of wood chips to provide the root system extra protection from very cold temperatures. It is important to continue protecting the root systems even if all the top growth gets killed after a particularly cold winter because new stems will rise from the base of the plant in spring, reaching up to 6 feet (1.8 m) by summertime.[4]

Fertilization and Growth: Since the plants like rich soil, a yearly application of compost around their root systems will encourage vigorous growth, in addition to keeping the roots from drying out during the hot parts of summer. Also, hold off on clearing away the protective mulch from the root systems and pruning back the dead stems until there is no danger of a late frost. We had a shrub survive many years that was buried in mulch each winter, and one year we uncovered the root system early in spring and pruned off the dead branches. Both of which stimulated growth, and then our gardens experienced a late frost that killed the unprotected plant down to the roots.

Cultivars: At the time of this publication, only one cultivar of chocolate berry is available in the nursery trade.

'Golden Lanterns': An attractive cultivar from England with bright yellow-green foliage that occasionally gets flushes of burgundy on new growth and along the leaf edges.

Related Species: The chocolate berry is closely related to honeysuckle (*Lonicera*) and snowberry (*Symphoricarpos*).

Propagation: There is not a lot of information available on propagating these wonderful plants, but we have read that the chocolate berry has naturalized by reseeding and has become an invasive pest in warmer areas, so one way to make more plants is to start the seeds indoors. We have also read that you can take hardwood cuttings from autumn to winter,[5] but we think that an easier way may be layering, by pushing down the lower outer branches, burying them in a small mound of soil, and securing them with a rock or landscape staple. This allows the plant to stay in place for a few seasons until separating from the mother plant.

Pests and Problems: We have not had any problems with this fruiting shrub, and the biggest concern with growing the chocolate berry in the northern parts of the United States is its cold hardiness.

Honeyberry

Lonicera caerulea, *L. caerulea* ssp. *edulis*, *L. caerulea* var. *emphyllocalyx*

Blue Honeysuckle, Fly Honeysuckle, Haskap

A widespread species inhabiting . . . the higher latitudes of the three northern continents. It has little or no merit for gardens, but has some botanical interest.

—W. J. Bean, 1914[1]

While the legendary botanist and plantsman W. J. Bean was not an enthusiastic fan of the honeyberry, we think it is a meritorious shrub that offers home gardeners several good attributes, such as a compact shape with loads of early flowers whose tart-sweet berries are the first fruits to ripen in the North. There are many species in the *Lonicera* genus, and most of the plants produce berries that are mildly poisonous. One of the few exceptions, however, is *L. caerulea*, commonly known as honeyberry. These are shrubby denizens of the cool boreal climates of the world and are found in the northern areas of China and Japan, as well as Siberia, the United States, and Canada.

Early-spring flowers.

Honeyberry bush.

Ripening berries.

Ripe berries.

Honeyberries have been harvested in their native lands for hundreds of years but have only recently come to the attention of North American gardeners and food growers. The blue honeyberry (*L. caerulea* ssp. *edulis*) and all of its cultivars are generally considered to be of Siberian or Canadian origin, and the plants generally produce fruit a week or two before strawberries ripen. In the 1950s, Soviet horticulturalists began breeding improved cultivars of these berries as a commercial crop, and for a while the only varieties available for home gardeners were the Russian cultivars.[2] In the 1980s, Japanese growers selected plants from the wild and began breeding this fruit, which ended up becoming a very popular (and expensive) market fruit in Japan.[3]

Honeyberry, along with its many varieties, is referred to as Haskap (*L. caerulea* var. *emphyllocalyx*), and these plants all originate from northern Japan and tend to ripen three to four weeks later than the Russian lineage.[4] Honeyberry is a very promiscuous and polymorphic plant (a species with many forms), which has resulted in lots of cross-pollination and interbreeding among the different plant subspecies.[5] Currently, all the mixed gene pool and complex alphabet soup of subspecies are no longer considered different species, but are recognized as different varieties of *L. caerulea*.[6] The main difference between the Japanese and Russian/Canadian types is when their flowers open and the fruits mature; they are referred to as early- or late-season types.

Honeyberries are the first fruit that ripens in our gardens, and although they are incorrectly marketed as being "like a blueberry," they are still a great early berry for fresh nibbling when no other fruit is ripe.

Growth Difficulty Rating: 1. Honeyberry bushes are low-maintenance shrubs that are easy to grow and have few problems.

Taste Profile and Uses: Honeyberries start off green then turn violet-purple and ripen a few weeks later with a blueberry-blue color that is often covered by a waxy bloom. Honeyberries reach their full size a month after flowering and require a total of about five weeks to mature and achieve a dark, rich color. The berries are at their best two weeks after they develop that deep blue color, and at this point they lose their astringency and turn sweet.[7] When the interior color of the berry is still green, the fruit has not fully ripened and can taste sour and bitter. The berry's average size is around ½ to 1 inch (1.3–2.5 cm) long, depending on the cultivar, and in shape can be oval, almost round, almost square, or a rounded teardrop. The flavor profile of this fruit is highly subjective, but the most common comparison is to blueberries, which we think is inaccurate. To us, the fruit has a pleasant sweet-tart flavor, with some cultivars having a slightly less acidic taste. The berries can be eaten fresh, put in smoothies, used in baking, or cooked down into preserves.

Ripening fruits are often hidden by the shrub's tendency to grow abundant foliage, and a gardener may have to gently shake bushes to loosen the ripe fruit or occasionally lift some of the shrub's branches for picking.

Plant Description: Honeyberry cultivars come in a wide range of sizes, which makes this shrub a good choice for gardens of every size. The height of the plants ranges from 3 to 8 feet (0.9–2.4 m) tall. The mature plant can grow into a rounded oval bush or grow into a vase form with the plant producing a large number of crowded branches. The leaves tend to be slightly pubescent (hairy) with edges that are fringed in ciliate (fine) hairs, and grow on the opposite sides of the stems. Due to extensive hybridization of plant cultivars, there is a range of leaf shapes, with some cultivars bearing long narrow leaves while other plants produce foliage that is more ovate.

Flowers: The flowers are arranged in pairs of pale yellow funnel-shaped blooms that hang upside down along the stems and often get produced in large quantities. The flowers emerge from the leaf axils and are lightly pubescent (hairy); they feature the curious appearance of being joined by a common ovary as a twin flower, which is not the case.[8]

Honeyberry is one of the first plants to bloom in spring, with flowers that are tough and cold-tolerant, and the plants in our gardens have produced fruit even after their blossoms got exposed to a temperature of 17°F (−8°C).

Pollination Requirements: Two compatible honeyberry varieties are needed for pollination, so purchasing two early-flowering varieties or two late-flowering varieties will ensure fruit-set. Most mail-order catalogs and nurseries that sell these plants list them separately by late or early cultivars and will occasionally sell them in sets for correct pollination.

Since the plants flower during the early part of spring while the weather is cool, the blossoms are reliant on bumblebees and flies for pollination. Some literature suggests that hand-pollination may be necessary for good crops, but we have not done this and our bushes produce reliable crops each season.

Site and Soil Conditions: Our bushes are planted in full sun, and during long, hot summers with little rain, the foliage has browned and shriveled up. This is probably the result of sun or wind scald, which many early-blooming honeyberry cultivars are prone to experience. To reduce the effects of heat-related stress on plants in the areas of the United States that are zone 6 (–10°F/–23°C) or warmer, these plants would benefit from a part-shade location that gets four to six hours of sun.[9]

Some sources suggest that honeyberry plants are not drought-tolerant and should be provided with supplemental water during long dry periods, which has also been our experience. The shrubs have shallow roots and benefit from a covering of leaf mulch or compost, which provides nutrients in addition to preventing the root systems from drying out during hot summer months. Also, honeyberry root systems prefer soil with an average level of consistent moisture but can develop disease problems if they are planted in an area that is constantly wet and soggy.[10]

Hardiness: Honeyberries are very tough plants and many of the Russian cultivars are hardy to zone 2 (–50°F/–46°C), although the majority of honeyberry plants are reliably hardy to zone 3, or –40°F (–40°C).

Fertilization and Growth: Honeyberries slowly expand and get wider through new shoots growing out of the root crown near the base of the plant. For many fruiting plants the age of the plant can affect its flowering cycle, and so it is with honeyberries, which tend to bloom earlier and grow larger fruit on older, established bushes.[11] They can be left to grow without being pruned for several years unless there is an excess of dead branches that need to be cleared. As the shrub matures, selective pruning may be necessary to remove less productive branches and to encourage new growth. Flowers appear on new stems, so try not to prune off the tips of branches—you will be removing a majority of the fruit.

Cultivars: As stated in the introduction, there is a lot of confusion about which varieties are compatible for pollination, so categorizing honeyberries as either early or late blooming is the best way to select the right plants for reliable fruit-set. There are more than a dozen different cultivars for each flowering type; we have listed a few below. You'll need at least two plants from one of the categories for pollination.

Early-Blooming Honeyberry Varieties (*L. carulea* ssp. *edulis*):

'Aurora': A hybrid between early- and late-blooming types that tends to bloom and ripen about a week after most Russian varieties on the market.

'Berry Blue': A vigorous and productive plant with an upright growth habit that grows to 8 feet (2.4 m) and produces abundant large, sweet, and tasty berries.

'Borealis': A University of Saskatchewan introduction that grows to around 4 feet (1.2 m) tall and produces good crops of sweet, large fruit.

Late-Blooming Honeyberries / Haskap varieties (*L. caerulea* var. *emphyllocalyx*):

'Blue Forest': A compact 2- to 3-foot (61–91 cm) tall cultivar that bears good crops of large honeyberries.

'Blue Hokkaido': Variety popular with Japanese growers that reaches 4 to 5 feet (1.2–1.5 m) tall with an upright growth habit and produces good-quality large, sweet-tart berries.

'Blue Pagoda': Another good Japanese cultivar that grows 4 to 5 feet (1.2–1.5) tall and produces large crops of quality fruit.

Related Species: There are over two hundred species in the honeysuckle genus, and plants such as mountain fly honeysuckle (*L. villosa*) and fly honeysuckle (*L. canadensis*) are close relatives.

Propagation: The most reliable way to generate new plants is by taking 5- to 7-inch (12.7–17.8 cm) long hardwood cuttings in the late fall that are about the diameter of a pencil and potting them up in soil with only a few buds above the soil line.

Pests and Problems: Honeyberries are relatively tough plants that are mostly pest-free, but they do have a few minor problems. Mildew and sunburn are common leaf maladies, and the degree to which they affect the plant will vary each season depending on environmental conditions, tending to be more severe in areas that are south of the Canadian border.[12] But neither of these problems seems to have a serious impact on the health of the shrubs that we grow, which still produce good harvests when these conditions are present on the leaves. However, both of these issues can make honeyberry bushes look ugly by midsummer, with curling and browning leaves.

The biggest pests are birds, which love the berries. Bushes with massive amounts of berries may need to be covered with netting to discourage flocks of winged fruit lovers.

Huckleberry

Gaylussacia baccata

Black Huckleberry

I use quotation marks around "huckleberries" deliberately because there is no universally acceptable berry that fits this word; it does not even cover a defined genus or any particular species within a genus.

—JENNIFER TREHANE, 2004[1]

As Trehane points out, there is no definitive "huckleberry" fruit. It's often confused with its close family relative the lowbush blueberry (*Vaccinium angustifolia*) because the two plants are often found growing right next to each other. At one point huckleberries were even included in the *Vaccinium* genus since they looked alike and grew in similar acidic soil environments. The big difference between these two North American native berries is their seed. If you eat a wild blueberry and its seeds are crunchy, then that's a huckleberry.

The huckleberry is an elusive fruit that straddles two different plant genus groups depending on your geographic location. On the West Coast it would be in the same family as blueberries and cranberries, as part of the *Vaccinium* clan. Along the East Coast of North America, scientifically inclined locals would tend to separate this plant into the lesser-known *Gaylussacia* genus, which has about fifty different species of flowering plants endemic to both South and North America. The real confusion begins when the two different plant groups intermingle in the Midwest and both *Vaccinium* and *Gaylussacia* begin sporting the name *huckleberry*.

For the purposes of this chapter, we are talking about only the most common species of huckleberry, also referred to as black huckleberry (*G. baccata*), which is endemic to eastern North America with a range that goes from Newfoundland across to Manitoba then down to the Great Lakes region, across the Midwest, to as far south as Georgia.[2]

In the early 1900s there was a common belief that the difficulties of domesticating the huckleberry would be similar to the blueberry, which was just beginning to gain attention as an agricultural crop. Because of the diversity of huckleberry species that were found growing in different environments, it was believed that the wild fruit could be improved by reducing the size and quantity of seeds, but this was never accomplished.[3]

Due to the size of the seeds, which are larger than those of their close relative the blueberry, these plants are commonly overlooked, which is a shame, because they have a wonderful flavor that is like a very rich-tasting blueberry with a sharp finish. We feel that if you can find a place to fit them into a garden, they could provide a low-maintenance fruiting crop that offers up delicious berries for late-summer pies and fruit preserves.

Growth Difficulty Rating: 1. Huckleberries are easy to grow if they are provided with acidic soil conditions and get consistent moisture.

Taste Profile and Uses: The berry-like drupe can range in size from ¼ to ¾ inch (0.6–1.9 cm). These berries start out light green and ripen to a deep purple-blue-black color, and are ready to eat from July to August. The interior is fleshy and sweet with ten edible, crunchy seeds. They have a wonderful, distinct flavor that is likened to a good blueberry, but the fruit contains more juice and has a sharper flavor. They have traditionally been gathered throughout New England as a wild crop used to make delicious fruit preserves and huckleberry pies.

The fruit was used by many Indigenous tribes for both food and medicinal uses. The Cherokee tribes chewed huckleberry leaves like medicinal tobacco to treat dysentery and tender gums.

Iroquois tribes smoked huckleberry leaves and drank a tea from the leaves to treat arthritis, colds, kidney ailments, rheumatism, tapeworm infections, and venereal diseases. Members of the Chippewa, Delaware, Mohegan, Menominee, Ojibwa, Potawatomi, and Shinnecock tribes drank huckleberry teas as a blood tonic and to treat colds, rheumatism, kidney ailments, fevers, and lumbago.[4]

Ripe and unripe berries.

Early-spring flowers.

Huckleberry flowers.

Ripe and unripe berries.

Plant Description: A small, densely branching deciduous shrub with a wiry habit that grows slowly by way of root suckers and forms into large colonies over time. The shrubs can have a variable shape, with their ultimate height determined by the amount of sun exposure. In open sunny areas the plants often grow only 1 foot (31 cm) tall, but when they are sited under the shade of a forest canopy, the bushes can reach 3 feet (91 cm). The leaves are alternately arranged and simple with smooth edges that have no serrated teeth. This is one of the key characteristics used to differentiate the huckleberry from the blueberry family, whose leaf edges are serrated. The elongated oval leaves grow 1 to 2 inches (2.5–5.1 cm) long, and both the upper shiny, dark green part of the leaf and the paler underside are covered with resinous golden dots—another key trait for identifying the plant.[5] Like their blueberry cousins, the leaves on huckleberry bushes turns shades of orange and crimson in the fall.

Flowers: Small, tubular, bell-shaped flowers grow ⅓ inch (9 mm) long and range in color from white-pink to scarlet-red, hanging in drooping panicles on the previous season's growth. The pedicles (stems) and calyces of the blossoms are covered with short fine hairs and yellow resinous dots, with the flowers arranged in clusters of three to seven blossoms.[6]

Pollination Requirements: The huckleberry is self-fertile, but there will be better fruit production with two or more bushes.[7] Many species of bees, as well as other insects, visit the flowers and help spread the pollen. Huckleberries can be found growing as part of the heath layer in pine barrens. Their flowers provide a critical source of nectar to the endangered Karner blue butterfly (*Plebejus melissa samuelis*).[8]

Site and Soil Conditions: Huckleberry is a fairly adaptable plant and does well in part shade, but will tolerate more sunlight if the root system gets consistent moisture. The shrubs are found in a diverse range of habitats, including conifer and mixed hardwood forests, meadows and fields, conifer bogs, and dry woodlands. We've seen large colonies of huckleberry growing as an understory plant in the shade of oak trees on Martha's Vineyard, as well as along the edge of a mixed hardwood forest clearing near our gardens. The best site for getting good fruit production would be one with enough sunlight to encourage fruit crops, with a small amount of shade so that the root systems will not dry out, such as four to six hours of light. The bushes are one of the few fruits that can grow in nutrient-poor soil with a preference for acidic conditions—a soil pH of 4.5 to 5.5. The plant's root system will tolerate both sandy and clay growing areas as long as they are well drained. We believe that the best result will be achieved with soil enriched with acidic organic amendments, such as broken-down oak leaves, or untreated pine mulch or bark. Our favorite acidic amendment is the "duff" (plant litter) beneath evergreen trees, which not only will acidify the soil but is more likely to contain living beneficial mycorrhizal fungi as part of its organic elements, and this root fungus is essential to the plant's well-being.[9] When digging up the soil from around evergreen trees, only scrape off a small amount of the topsoil from each tree to avoid damaging the roots. We use a thick mulch of pine needles to suppress weeds, and the needles will also provide nutrients to the soil as they break down over time.

Hardiness: Huckleberry is cold-hardy to zone 4, or to −30°F (−34°C).

Fertilization and Growth: Huckleberries thrive in acidic conditions, so top-dressing the areas around the roots with pine needles, pine sawdust, or pine bark will prevent the roots from drying out in the heat of summer while also adding beneficial nutrients to the root systems as they decompose over time.

Cultivars: At the time of this publication, there are no named cultivars of the huckleberry.

Related Species: The blue huckleberry, also known as dangleberry (*G. frondosa*), has edible fruit, with a native range from New Hampshire to Florida, across to Ohio and Louisiana. Dwarf huckleberry (*G. dumosa*) is a low-growing ground cover that can reach around 18 inches (45.7 cm) tall, and produces berries that are also edible.

Propagation: The plants are difficult to grow from seeds; it is much easier to propagate new shrubs by digging up root suckers in spring while the plant is still dormant. Pot up your divisions using some of the soil from the plants' original growing area to bring along the mycorrhizal fungi, as well as to reduce the effects of transplant shock. Grow the plants in shade for two to three seasons to increase the size of their slow-growing root system, before planting them in the ground.

Pests and Problems: Although there are few major pest or disease problems for the huckleberry, it is an important source of food for songbirds, turkeys, game birds, and other wildlife. So, protecting your fruit much as you do blueberries will be necessary if you want to harvest some for yourself.

Jujube

Ziziphus jujuba

Chinese Date, Red Date

> *Chinese dates, or Jujubes, are one of the earliest tree crops to be cultivated by mankind. Here in America, they have been quietly sitting in the agricultural background for over 150 years.*
>
> —ROGER AND SHIRLEY MEYER, 1994[1]

Roger and Shirley Meyer were among the many American fruit pioneers who attempted to introduce the jujube (*Ziziphus jujuba*) to America, when they imported jujube cultivars from the Nanjing Botanical Gardens in the 1990s and created a home orchard to sell fresh fruits in California at farmers markets. But there have been many food explorers that tried to establish these plants as a fruit crop in the United States. Over a century ago in 1837, Robert Chisolm was the first to import jujube seeds from China to the US, planting them in Beaufort, North Carolina,[2] but it was the great food explorer Frank Meyer (no relation to Roger and Shirley) who did the most to introduce the jujube to the United States.[3] Working on behalf of the USDA,

F. Meyer made four exploration trips to China from 1905 to 1918 and collected thousands of useful plants, including over sixty-four types of jujubes in the form of seeds or scion wood. He sent these back to the United States for planting in the USDA Plant Introduction Station in Chico, California.[4] Unfortunately, the Chico orchard was abandoned and many of the early promising cultivars were lost, although a few of Frank Meyer's best varieties, like 'Li' and 'So', are still being cultivated in American gardens. The most recent pomological pioneer that has tried to establish the jujube in the United States is Shengrui Yao, a pomology professor at New Mexico State University. In 2011 Professor Yao imported thirty-four cultivars from China that were planted in four university test growing plots to rate their viability as a potential crop. These new trees are now being sold as the trademarked 'AmeriZao' jujube varieties.[5]

Our jujube trees are healthy and doing well, even though there is not a lot of information available on how these trees perform in the northern parts of the United States. Although jujube has done well in the southern parts of America, this is one example in this book where we are experimenting with these trees to see if they will fruit. So if you would like to join the earlier fruit pioneers trying to establish the jujube in America, these fruits might provide you with a new taste, one that has been loved in China for more than four thousand years.

Growth Difficulty Rating: 1. Jujube trees are trouble-free plants that have few pests but must be planted in full sun and require a long, hot growing season to fully ripen fruit.

Taste Profile and Uses: The jujube can come in a wide range of shapes and sizes depending on the variety, and trees can produce fruit that is round, oval, pear-shaped, or an elongated spherical shape. The fruit is a ½- to 2-inch (1.3–5.1 cm) long drupe that contains one long, thin seed. Jujubes start out as a dense, firm fruit covered by a thin, edible skin with a light green color, which develops mottled chestnut-brown patterns that slowly cover the fruit until it is completely (or almost completely) brown. This happens from mid- to late autumn, and at this stage the fruit can be eaten and has a firm texture and a flavor reminiscent of an apple. Several days after picking, the fruit begins to slowly dry out, and over time the outer skin begins to wrinkle and the cellular structure of the fruit becomes starchy or date-like. In its best form, the fruit's texture turns pleasantly chewy with a mild date flavor. It is used for a whole range of traditional Chinese cooking, from sweet to savory.

A young 'Sugar Cane' tree.

Jujube flowers and foliage.

Jujube flowers opening.

Small fruits forming.

Jujubes are often pitted and added to stir-fry dishes, soups, and rice, and used like a dried fruit in sweet baked breads. We have purchased dried jujubes in Asian supermarkets, where they are sold as a dried snack food, and they are also popular as a dessert tea when combined with honey. There is also a long history in China of jujube fruit and seeds being used in traditional Chinese medicine.[6]

Plant Description: Jujube is an attractive tree that grows 20 to 25 feet (6.1–7.6 m) in height, although grafted varieties usually grow 12 to 20 feet (3.7–6.1 m) tall. The 1- to 2½-inch (2.5–6.4 cm) long leaves are an attractive, glossy bright green color with an oval to lance shape and finely toothed margins. The tree branches can grow with a natural zigzag pattern. The limbs produce some thorns, ranging from branches that are covered with them to some cultivars that are close to thornless. Mature trees frequently grow with a slightly contorted shape or an attractive weeping habit.

Young fruit ripening.

Flowers: Tiny flowers are ¼ inch (6 mm) in diameter and are a light greenish yellow color; they are pollinated by various types of small flying insects. The flowers grow out of the leaf axils on the current season's growth, produced in groups of up to thirteen blossoms, and the trees produce flowers over a long period from late spring into summer.

Pollination Requirements: Jujube trees generally require two trees for cross-pollination. There are a few partially self-fertile varieties such as 'Li', but even the self-fertile trees will produce much better yields with two trees planted together.

For many years it was assumed that all jujube trees would be compatible for pollination, but recently Professor Shengrui Yao and her colleagues have conducted flower studies of different cultivars and determined that there may be two distinct lines of cultivars whose flowers open at slightly different times of the day.[7] Currently most of the different jujube cultivars are all being sold as being able to cross-pollinate with one another.

Site and Soil Conditions: These trees love as much sun exposure as possible, and in the northern parts of the United States the jujube should be planted only in full sunlight since their fruit will require a long, hot growing season.[8] The trees are adaptable and have been grown in a wide range of soils; they do well as long as the root systems are in well-drained soil. After getting established for a few growing seasons with consistent watering, they have thrived in the dry desert condition of the Southwest, but should occasionally receive supplemental water throughout a growing season for the best-quality fruit.[9]

Our trees are planted in well-drained soil with a heavy clay content that receives consistent rain (or snow) throughout the year, and they have had no problems. For a long time it was assumed that the plants would thrive only in California, Texas, the Southwest, and the Deep South. But over time jujube cultivation has spread to other parts of the United States; trees have been planted in Maryland, Ohio, Pennsylvania, and Washington, DC.[10] We have seen established, full-sized trees with large fruit crops in in the Brooklyn Botanic Garden, so the growing range for these trees is greater than was first understood.

Hardiness: The trees are rated as hardy to zone 5, or −20°F (−29°C)—and there are records of jujubes that have survived winter temperatures of −22°F (−30°C)[11]—but in the colder parts of zone 5, the main problem may not be the cold but the length of the growing season. Jujube fruits require a long, hot growing season for the fruits to mature, and in the colder parts of zone 5, there might not be a sufficient number of hot days. This is why many nurseries list the jujube as being hardy to zone 6 (−10°F/−23°C). But if you are in zone 5 it might still be worth experimenting with the trees, though they should be planted in an area of maximum sun exposure.

Fertilization and Growth: The trees seem to grow well in average soil conditions and have thrived with neglect, but an annual application of well-aged manure or compost will make them more vigorous. The root systems of jujubes freely produce suckers that grow around the base of the tree. These should be removed at the soil line with pruners or eliminated with a mower to keep the ground around the trees clear, and to ensure that the main trunk of the grafted cultivar receives all of the dominant growth.

Cultivars: There are about eight hundred named cultivars of jujube in China,[12] but in the United States the choices are limited—about twenty to thirty grafted varieties are available through specialty mail-order nurseries. Here are a few of the varieties that have a reputation for producing good-quality fruit.

'CoCo': A more recent Ukrainian cultivar from the Main Botanic Garden in Kiev, which produces fruits that a have distinctive coconut-like flavor.

'Lang': A Frank Meyer introduction that produces very large, pear-shaped fruit on a tree that is almost thornless.

'Li': Another Frank Meyer introduction whose high-quality fruits are medium to extra-large in size.

Related Species: There is a species of jujube that is native to India (*Ziziphus mauritiana*).

Propagation: Jujube trees can be grown from seed, but the most common way to propagate new plants is by grafting. The root suckers that form around older, established trees can be carefully dug out with some roots attached, and potted up in soil. These can be grown out as new trees or used as a rootstock to graft cultivated varieties and create new jujube trees. Just be careful to dig up suckers that are far enough away from the trunk of the tree so that you do not cause any damage.

Pests and Problems: Jujube trees are fairly pest and disease free in the United States, but birds have devoured ripening fruit growing in California orchards, so crops may need some bird protection.

Juneberry

Amelanchier spp.

Saskatoon, Serviceberry, Shadblow, Shadbush

> *...Juneberries have been a source of food-supply to explorers, prospectors and pioneers, who testify to their value as nourishing esculents and pleasing dessert fruits. Juneberries are as yet little used where they must compete with other fruits, although they have many qualities to commend them for domestication.*
>
> —L. H. BAILEY, 1930[1]

Although people have been eating juneberries for thousands of years, they are still relatively unknown as a food crop in modern American culture. There are about twenty to twenty-five species of juneberries (*Amelanchier*) growing in almost every part of North America: in open plains, forests, dry woods, mountain regions, sheltered slopes, wood margins, conifer forests, and pine barrens. The various species grow across wide areas that overlap, and their range extends from Alaska across the bottom of Canada, south through the Midwest and East Coast, down to Texas and Florida.[2]

Juneberries were widely used by Indigenous tribes as an important food source and an ingredient in pemmican, a food that is a blend of berries, dried meat, and fat.[3] Pemmican became a way to preserve food that nourished tribes through long winters, and the use of juneberries as a survival food was passed on to the explorers who traveled across the United States. Lewis and Clark, for instance, used the small fruits as a survival food as they traveled west across the Plains.[4]

In our neck of the woods in upstate New York, juneberries are traditionally referred to as *shadblow* or *shadbush*, because they produce early-spring flowers that coincide with the seasonal migration of shad fish swimming up the Hudson River. These names are specific to the *Amelanchier* species (*A. laevis* and *A. canadensis*) growing along the East Coast in forest areas. In western North America the name *saskatoon* (*A. alnifolia*) is a reference to the bush form of juneberry that grows in sunny prairies and the Canadian provinces above California and Washington. These plants are the juneberries most commonly cultivated for commercial fruit production, and sold in the nursery trade as fruiting shrubs. But unfortunately, the plants that nurseries sell are often mislabeled, because many *Amelanchier* species are difficult to distinguish. This only becomes more challenging given the different species have been crossed with one another to produce numerous hybrids.[5]

Flower buds.

A juneberry tree or bush could be a great addition to any garden or backyard for its delicious fruit and stunning displays of early-spring flowers, which are a welcome sight after a long winter.

Growth Difficulty Rating: 2. Juneberry is adaptable and fairly easy to grow, although the plants and fruits are susceptible to several pests and diseases.

Taste Profile and Uses: The fruit looks most similar to a blueberry in appearance, but it is technically not a berry but a pome, which ranges in size from 3⁄8 to 5⁄8 inch (1–1.6 cm), depending on the species or cultivar. They start out green and develop a range of colors from fuchsia, to maroon-red, to dark purple and almost black, overlaid with a waxy bloom. The flesh is soft and contains a few edible seeds that taste pleasantly similar to almond. Although some sources liken the flavor of juneberries to the taste of a blueberry, this is not quite correct. They have a unique flavor that combines several different popular fruits including dark cherries, blueberries, apples, and raisins, with a finish of almonds. Some people may find that juneberry is sweet but bland because the fruit is very low in acidity and lacks the sour notes that give many good fruits a balanced flavor. We think that this is one case where the seeds actually make the fruit to taste better. The fruits are an excellent source of iron and vitamins A and E, and have high levels of protein, magnesium, calcium, potassium, and antioxidants.[6]

Juneberries have a short picking season and reach peak flavor over a few weeks, and the best way to enjoy this fruit is fresh off the branch. The fruit can be used as a substitute for any baked item that requires a blueberry. Juneberry pies are amazing.

Plant Description: All of the plants in the *Amelanchier* genus make handsome garden additions. You'll find a variety of plant shapes and sizes depending on the species and their cultivars, ranging from bushes that are about 4 feet (1.2 m) high to trees that can be 25 feet (7.6 m) or taller. Juneberry stems produce simple, alternately arranged leaves that are round-oval to oblong-oval, with finely toothed edges. Individual species can grow leaves and flower buds that are tomentose (hairy), or grow leaves that are completely glabrous (smooth). Juneberry trees and shrubs display magnificent fall foliage in a whole range of colors including yellow, orange, and crimson red.

Flowers: Juneberry trees feature attractive star-shaped white flowers with five narrow petals that range in size from 1⁄2 to 11⁄4 inches

(1.3–3.2 cm) long. These form on pendulous racemes (flower clusters) that are 2 to 4 inches (5.1–10.2 cm) in length and open in early spring when the leaves have not completely emerged from their leaf buds. Although the flowers are small, a mature juneberry tree in full sun can produce spectacular displays of blossoms that have many of the best decorative attributes of a flowering cherry tree.

Pollination Requirements: Juneberries are self-fertile but will produce more fruit with another tree or bush planted nearby.

Site and Soil Conditions: Juneberry plants tolerate a wide range of growing conditions, from coarse sandy soil to poor-quality dirt with heavy clay content, and they are adaptable to many different planting areas as long as the soil is well drained. Because the trees are found growing in shady forest areas, many sources list them as being able to grow in locations with a lot of shade, but in our gardens *A. alnifolia* and *A. canadensis* were constantly plagued by disease problems when planted in part shade. Individual parts of the United States may have less disease exposure, but we feel that for fruit production, juneberry trees should be planted in full sun.

The plants in this genus are often mixed up, and there are many misleading assumptions about growing juneberry. In general, the species that are moderate-sized shrubs are mostly native to the western part of the United States and Canada, where they are found growing in full sun. The species endemic to eastern North America tend to grow into multi-trunked, medium-sized trees that can flower in shaded conditions but will ultimately produce larger crops and have fewer pest problems in full sun.

Hardiness: Juneberry hardiness varies with different species and is a reflection of the diverse geographic range of individual species. *A. arborea, A. alnifolia, A. laevis,* and *A.* × *grandiflora* are rated as being hardy to zone 4, or −30°F (−34°C); *A. canadensis* is rated at zone 3, or −40°F (−40°C).

Fertilization and Growth: Juneberries are adaptable plants that grow in many types of soil conditions, and while they will survive with little effort, a gardener will be rewarded with better fruit crops by enriching the soil with a thick yearly application of compost. Some types of juneberry may be prone to suckering; you can prune these into a neater plant with three to four trunks if you wish.

Cultivars: With so many species and hybrids between the species, it is difficult to know which plants to select for a particular site. We feel some of the best juneberry cultivars are from the cultivated variety *A.* × *grandiflora,* which is a cross between *A. laevis* and

Lovely spring flowers.

Tree in bloom.

Juneberries ripening.

Juneberry harvest.

A. arborea, and there are many cultivars that resulted from this hybridization in the nursery trade. We planted the 'Autumn Brilliance' cultivar of *A.* × *grandiflora*, which lived up to its name by producing outstanding fall color displays in addition to good crops of flavorful fruit.

'Autumn Brilliance': A lovely shade tree that grows to 25 feet (7.6 m), produces good-quality fruit, and is hardy to −35°F (−37°C). The cultivar is fairly pest-free, but we do get small amounts of cedar rust each year.

'Regent' (*A. alnifolia*): A compact, early-flowering cultivar that grows 4 to 6 feet (1.2–1.8 m) tall, with extra-sweet fruit. It is hardy to zone 4 (−30°F/−34°C) and has good disease resistance.

'Thiessen' (*A. alnifolia*): Introduced in Saskatchewan in 1978, this cultivar is a large bush that grows 12 to 15 feet (3.7–4.6 m) high and yields the largest fruit of all the juneberry varieties.

Related Species: Juneberries are related to hawthorns (*Crataegus* spp.) and mountain ash (*Sorbus* spp).

Propagation: Juneberry cultivars that produce suckers near the base can be dug out of the ground with a small section of root attached to the bottom of the suckering branch. This is best accomplished in late winter to early spring, before the plants start flowering. You can also start juneberries from seed that has been subjected to cold stratification; however, up to a third of the seedling plants will end up being different from the parent plant.[7]

Pests and Problems: Rust (*Gymnosporangium* spp.), leaf spot (*Entomosporium* spp.), fire blight (*Erwinia amylovora*), and powdery mildew are all problems that occasionally affect juneberry plants, particularly in seasons that are excessively rainy. We feel that the high humidity of the Northeast can lead to problems with powdery mildew and fungal diseases; planting the trees in full sun may help to alleviate these problems. Discarding any of the diseased parts of the tree will also help to keep the plant healthy. But by far, the biggest pests are birds, which will gladly strip the fruit off trees before it's fully ripe. Still, almost-ripe juneberries that have turned to a deep scarlet-red are good enough to use for making pies, even if they do not have the sweetest flavor. Sometimes it is better to pick a large harvest of underripe berries for baking, instead of waiting a week to get a small harvest of fully ripe fruit left by the birds.

Korean Stone Pine

Pinus koraiensis

Cedar Pine, Chinese Pine Nut,
Korean Nut Pine,
Korean White Pine

> *The conifers possess one particular advantage in the fact that they seem to require no cultivation at all. Many of them are lovers of mountain sides and of rough land and may be grown where other fruit trees would not thrive, or at least could not be raised profitably.*
>
> —Robert T. Morris, 1921[1]

At one time, the Korean pine (*Pinus koraiensis*) was a common conifer tree native to thousands of miles of wild forests in the Russian Far East, northeast China, and North Korea, with smaller populations growing in South Korea and Japan. These stately trees grow into 100-foot (30.5 m) giant pines that filled the forests. But over the past two decades in countries like China and Russia, this species has been reduced by more than 50 percent through extensive legal and illegal logging. The tree's lumber is extensively used in the

New growth.

Needles showing silver-blue undersides.

Young Korean stone pine.

A handful of Korean stone pine nuts in the shell.

construction industry, which has reduced the populations of these beautiful trees to just a fraction of their original numbers.[2]

To most observers, the Korean pine tree has a similar habit and appearance to the American white pine (*P. strobus*), and at a distance the trees would be difficult to tell apart. From a culinary standpoint the difference between the two species is the wonderful seeds that Korean pine trees produce. The seeds are about the size of a small pistachio and are better known as *pignoli* or "pine nuts"; they have a delicious, rich, starchy flavor that compares favorably with the better-known Italian pine nuts. There is a common misconception that all of the imported nuts sold in the US food markets are produced from Italian stone pine trees (*P. pinea*), grown in Italy. But in actuality, a large portion of the pine nuts sold in the US and internationally are harvested from Korean stone pine trees that are grown in China.[3] These nuts are usually sold at a high price because the majority are harvested by hand off the wild trees using long, hooked poles to dislodge the cones, so collecting is labor-intensive and time-consuming. Plus the trees tend to produce heavy crops every three to five years with lean harvests in between, so the nuts are unstable as a market crop.[4]

If you love pine trees and have a space to grow two large ones, Korean pines would be an attractive addition to a large garden or orchard area, and they could provide harvests of delicious edible nuts.

Growth Difficulty Rating: 1. Korean stone pines are easy to grow with few major problems.

Taste Profile and Uses: Edible Korean pine nuts are produced in the female (seed-bearing) cones. The pinecones develop a circular shape comprising a series of scales. They contain two ovules that will develop into pine nuts if they have been pollinated. The cones develop over a two-year period that begins with young cones being produced in the spring of the first year. Then after winter in the second year, the green cones are pollinated and the seeds (pine nuts) inside start to develop. These will be fully mature by the end of the year in autumn, when the cones begin to turn brown and crack open to release the pine nuts.[5]

Gloves should be used when harvesting the nuts because the cones that encase the seeds are covered with a very sticky resin. Pick the cones off the tree toward the end of the second season when the scales are still closed, and store them in bags to allow the cones to finish ripening the seeds. The cones will dry out and the individual scales will crack open to reveal the nuts encased in brown papery shells, which can be harvested out of the cones. Pine nuts kept in their shells will stay fresh for several years, but after extraction the raw nuts should be refrigerated or frozen, so that their delicate flavor doesn't spoil. Pine nuts can be consumed raw but have a richer flavor when they are lightly toasted. The nuts are fiber- and protein-rich.[6] They are great for both sweet and savory recipes and are commonly used for making pesto or cookies.

Some sources say the trees are mature enough to produce pine nuts when they reach 6 to 8 feet (1.8–2.4 m) in height, but we have not seen this. After five years our trees grew to about 10 feet (3.1 m) before one of them produced a single cone that was stolen by wildlife.

Plant Description: The Korean stone pine is a handsome conifer with evergreen foliage that grows into a pyramidal outline in youth and matures into a rounded shape. Under cultivation the trees reach a common height of around 30 to 50 feet (9.1–15.2 m), although there are individual wild trees that have grown to 100 feet (30.5 m). Its dark green needles have silver-blue undersides; they grow 2½ to 6 inches (6.4–15.2 cm) long and are borne in bundles of five at the end of a short shoot, living for three years before falling off. The female cones are 3½ to 5 inches (8.9–12.7 cm) long with a green color that matures to brown. The tiny oval male pollen cones are orange-brown and are produced in clusters on the base of new branch shoots.

Pollination Requirements: The trees are self-fertile but will consistently produce larger crops when two trees are planted together. Pine trees rely on the wind to spread the fine, dust-like pollen from the male pollen cones to the female cones.

Site and Soil Conditions: These evergreens prefer a growing site with at least eight hours of sunlight and good drainage. They are adaptable trees that will grow in many different types of soil such as sandy loam or clay soil. Since the tree is a conifer, it will do best in acidic soil, but the plant's root systems will tolerate growing areas with a heavy clay content as long as the ground is enriched with acidic organic amendments such as broken-down oak leaves, or untreated pine mulch or bark. Our favorite acidic amendment is the duff (plant litter) beneath evergreen trees, which will not only acidify the soil but is more likely to contain living beneficial mycorrhizal fungi as part of its organic elements. When you dig up the soil from around evergreens, scrape off only a small amount of the topsoil from each tree to avoid damaging the roots.

Hardiness: Of all the pines that produce edible nuts, this is one of the hardiest trees and can survive in zone 4, or −30°F (−34°C). But there are reliable horticultural sources that list it as potentially able to survive even colder temperatures.[7]

Fertilization and Growth: Korean pines exhibit slow growth during the first years of their lives. After their root systems get established, the trees can grow at a faster rate; the pines in our gardens have produced 8 to 14 inches (20.3–35.6 cm) of healthy growth each season. Since pine needles are cast off every three years, the trees can slowly create an acidic growing environment in the ground that they inhabit. Providing the trees with a thick mulch of pine needles or pine bark chips will help create an acidic environment for their root systems and keep them moist and cool for the first few growing seasons after planting, until the tree matures enough to produce its own natural layer of pine mulch carpeting.

Cultivars: There are no cultivars of Korean pines that have been selected for nut production. Because these are handsome

Needles of Korean stone pine.

ornamental trees, all the cultivars that have been introduced by nurseries are for ornamental characteristics such as their growth habits and decorative foliage.

'Glauca': Cultivar that produces attractive, soft-textured needles with a pronounced blue tint and is hardy to zone 4.
'Oculis Draconis': A handsome form of the tree with variegated yellow stripes across the foliage that is hardy to zone 3.
'Silveray': A fastigiate (upright-columnar) form of the tree with twisted silver-bluish needles introduced in the Netherlands in 1978. It is hardy to zone 4.

Related Species: This species is related to several other pine trees that produce delicious edible nuts such as the Siberian stone pine (*P. sibirica*),which is hardy to zone 1, and the Italian stone pine (*P. pinea*), which is hardy to zone 7.

Propagation: Trees can be started by fresh seeds, which can be directly sown in pots, but germination will be improved with cold stratification (exposure to cold temperatures) by storing the seeds in a refrigerator for three to seven weeks prior to planting. Korean pine trees are also traditionally grafted onto another pine tree as a rootstock.[8]

Pests and Problems: We haven't had any pest problems with our trees, and this species is widely listed as being resistant to problems such as white pine blister rust (*Cronartium ribicola*). After the trees begin to produce large crops of cones filled with nuts, it's best to gather them in the late fall while they are still green rather than waiting until they drop to the ground and get taken by birds and squirrels.

Ecological Importance: The most important aspect of the Korean stone pine's biology is its important ecological relationship to wildlife species in its native lands. During the wintertime the pine nuts provide the critical food source that allows a wide range of animal species to survive through a long winter, including deer and wild boar. Those two animals are the primary winter food hunted by one of the most majestic animal species on Earth, the Amur tiger (*Panthera tigris* ssp. *altaica*), which is native to a small area in Russia and is critically endangered. Because large stands of Korean stone pines are harvested as lumber for building, those forests are dwindling. The Russian government finally imposed a ban on the illegal logging of Korean stone pines in 2010 to help in the conservation of the trees, which are the key to the Amur tiger's survival.[9]

Lingonberry

Vaccinium vitis-idaea

Cowberry, Foxberry, Mountain Cranberry, Partridgeberry

> *Throughout the whole of N. Canada, hunters and trappers, as well as the native Indians, have frequently depended on it for food. It is valuable for the shrubbery border, where the strong contrast of the dark green foliage and the bright colored persistent fr. is very striking.*
>
> —L. H. BAILEY, 1930[1]

If you go inside any of the international IKEA retail outlets, you will encounter a display of lingonberry jam, prominently presented among all of the other retail items. This is because the lingonberry is Sweden's national fruit and viewed as a country-wide treasure that is an essential part of many traditional meals. The small, tart scarlet-red fruits are still commercially harvested by hand like they were centuries ago, and are also very popular throughout all of the Scandinavian countries that they are native to.[2] During the German occupation of Norway in World War II, food rationing forced hungry Norwegians to forage for wild lingonberry fruits, which were harvested so heavily that inspectors, who were referred to as the "Lingonberry Police," were hired to impose fines and keep people away from the wild fruits until they were ripe and to make sure that the lingonberries were equally distributed among the citizenry.[3] The love for this little berry was not forgotten when Scandinavian immigrants settled in the American Midwest in the late nineteenth century. At one point Minneapolis and the surrounding areas purchased eighty-six hundred barrels of lingonberry from Newfoundland, Canada, and that was for just one year![4]

Lingonberry plants are now botanically divided into two separate types of bushes. The first is often referred to as the cultivated European type (*Vaccinium vitis-idaea*), which is distributed across a large geographic range of arctic and sub-arctic countries including Canada, Greenland, Scandinavia, northern Europe, Alaska, Asia, and the northern parts of the United States in mountainous areas.[5] The European type grows in a wide range of habitats, and most of the improved fruiting varieties have been produced from these plants. They also have another distinctive feature, which is that they generate two separate flower blooms and two distinct harvests each season.[6]

The second type of lingonberry is referred to as Dwarf lingonberry (*V. vitis-idaea* ssp. *minus*), which is occasionally listed as the American lingonberry or simply referred to as a subspecies of the European lingonberry. Regardless of its confusing, extra-long Latin name, this is a lower-growing plant that produces a single crop per season and is found in many of the same areas that the European lingonberry grows. These low-growing, trouble-free plants can provide food and decorative color, and are easy to tuck into smaller spaces within a garden.

Growth Difficulty Rating: 1. Lingonberries are easy to grow with few major problems in any garden setting that receives a consistent amount of moisture.

Taste Profile and Uses: The berries are ¼ inch (6 mm) or slightly larger in size with pinkish red to deep red fruit, arranged in clusters that hang off short stalks. The raw fruits have soft flesh similar to their close relative the cranberry, but the lingonberry has a less astringent taste and is more palatable. When the berries are cooked with a sweetener, the result is a rich, tart cranberry-like flavor that is wonderful for jams, jellies, juices, or as a substitute for any recipe that calls for cranberries. Because of their tiny size, picking the berries can be a laborious task; people harvesting the fruits use handheld berry rakes to expedite this process.

Lingonberries can persist on the shrub for a long period, from the fall through winter, and actually become sweeter after an exposure to frost, but they may be snatched by hungry wildlife if you leave them on the plants for too long. The fruits can be stored in a refrigerator for several weeks because lingonberries contain high levels of benzoic acid, which is a natural preservative, or they can be frozen. The berries have both vitamins A and C and contain four times the antioxidants of blueberries,[7] which is why they are used in various medicinal preparations, including cough syrups.[8] Lingonberries are used in traditional folk medicines to treat blood disorders, inflammatory diseases, wounds, gastric disorders, rheumatism, and urinary tract infections.[9]

Plant Description: The plants are low-spreading evergreen shrubs that grow 3 to 16 inches (7.6–40.6 cm) tall with creeping branches. The plants form into dense colonies through spreading fibrous root systems that grow just beneath the soil, then send up new shoots that continue to expand their width. They can also expand through stolons, which are plant stems that hit the ground, begin to grow roots, and form into new shrubs. New branches begin as green shoots, then develop into a mature burgundy-brown color and take on a wiry appearance as they age. The low ground-cover-like shrubs feature attractive oval-shaped leaves with a shiny bright to dark green color and a leathery texture, that are ¼ to ¾ inch (0.6–1.9 cm) long. The leaves are alternately attached to the stems, and their edges are occasionally slightly curled. Cold fall temperatures turn the foliage on lingonberry plants to a bronzy-purple color, which adds an attractive feature to the plants throughout the winter season.

A ripe berry amid foliage.

Flower buds.

The delicate, bell-shaped flowers.

Ripening fruits.

Lingonberry fruits.

Flowers: The plants produce tiny, whitish pink bell-shaped flowers with four petals that are 1⁄4 inch (6 mm) long and hang in clusters at the end of branch tips. European lingonberry plants often produce flowers at two distinct times during the growing season, one in the middle of summer and the other toward the early fall, but this is dependent upon a series of different environmental factors such as warm spring temperatures, as well as the geographic location of the plants. The American dwarf lingonberry flowers just once a year in spring.

Pollination Requirements: The plants are self-fertile, but because almost all the plants sold in the nursery trade are very small, it is best to plant several specimens in one location to fill out a planting area and to get a larger quantity of berries. The flowers will also benefit from cross-pollination of two or more plants, and lingonberry pollen attracts a large number of native pollinating insects.[10]

Site and Soil Conditions: They prefer well-drained acidic soil conditions but are adaptable plants in their native haunts and are found in the wild in both sunny and shady settings; growing in bogs, on open tundra grasslands, and along rocky slopes.[11] Lingonberries prefer soil amended with acidic organic matter. If you live in a warmer growing region such as zone 6 (−10°F/−23°C) or 7 (0°F/−18°C), the plants should be sited in a part-shade location that gets sun in the cooler parts of the day, such as the early morning or late afternoon. But in the colder regions of the United States—zone 5 (−20°F/−23°C) or colder—lingonberries can be planted in full sun. Either way these plants' root systems do not do well in waterlogged conditions, and they should be weeded regularly since their shallow roots do not like competition.

Hardiness: Lingonberries are denizens of the boreal forest and very tough plants, surviving brutal winter temperatures that plummet down to −50°F (−46°C), or zone 2! Because these plants are native to the arctic, they have a preference for cool summer temperatures and will not grow well in the warmer areas of the United States that are above zone 7 (0°F/−18°C).[12]

Fertilization and Growth: Lingonberry stems can take a few years to begin fruiting, and these branches will continue to yield flowers and berries for about five seasons, at which point they begin to decline and are less productive. If you prune the older stems off the plants, this will encourage fruitful young branches to replace them and promote better harvests. When pruning, leave at least six to eight vigorous branches on each bush; this should allow enough space for new growth and open the plants up for more air circulation.

Lingonberries grow on shallow root systems so it's important to water the plants for the first few years until they are established, and clear weeds away to reduce the competition for moisture and nutrients. In early spring a thick mulch over the plants will help reduce weeds and protect the shallow roots from drying out. A mulch that combines an acidic soil medium, such as pine needles or pine duff (pine litter and needles) mixed with sand, will help to acidify the soil and is more likely to contain living beneficial mycorrhizal fungi as part of the soil's organic elements.

Cultivars: There has been an interest in the cultivation of lingonberry as an evergreen ground cover because the plants are small and adaptable to many garden locations. Over the past ten years or so, lingonberries have become more popular in the nursery trade. About a dozen varieties are now available, selected from wild plants that had promising attributes such as large berries and attractive foliage.

'Erntedank': A cultivar that produces heavy yields of small to medium-sized berries that was cultivated from wild plants growing in Germany.

'Koralle': A cultivar from Holland that was introduced in 1969. Initially touted as an ornamental ground cover, it is now the leading variety grown for its berries. It blooms in midsummer and in late summer, producing a crop in mid-August and a second batch in October.

'Red Pearl': A popular, fast-growing Dutch cultivar that produces slightly larger berries but is less productive than other cultivars. It is tolerant of bad soil conditions and has two flowering periods.

Related Species: The lingonberry belongs to the *Vaccinium* genus, a large group of plants that includes the blueberry (*V. angustifolium*) and cranberry (*V. macrocarpon*).

Propagation: Stem cuttings taken in late summer and overwintered can be planted out after a few years when they have developed good root systems. Digging up and dividing young, rooted stolons is another easy way to generate new plants.

Pests and Problems: Lingonberries have very few pest problems, but the berries grow low to the ground and may be eaten by birds and other small animals.

Mayapple

Podophyllum peltatum

American Mandrake, Hog Apple, Indian Apple, Raccoon Berry, Wild Lemon

> *The fully ripe fruit of the May-Apple . . . is familiar to every country boy of the regions where the plant abounds and, although Asa Gray described it as "mawkish, eaten by pigs and boys," in its fresh state it has a peculiar flavor very agreeable to most human grown-ups.*
>
> —Fernald and Kinsey, 1958[1]

The mayapple is a strange, leafy inhabitant of the shady woodlands of eastern North America with a range from Ontario and Quebec south to Texas. Mayapples form giant colonies around the feet of deciduous forest trees, and greet the warm early-spring air by pushing up through the forest leaf litter as the stalks unfurl into giant shredded green umbrellas that can carpet the areas between the trees. Mayapples are one of the most distinctive of the early-spring wildflowers of the eastern forests, and their attractive, nodding spring flowers and large, umbrella-shaped leaves are one of the clear signs that spring has arrived. Such a distinctive plant with unusual leaves did not fail to catch the attention of the first Europeans that arrived in North America. In 1616 the French explorer Samuel de Champlain saw members of the Huron-Wendat tribes eating mayapple fruits; he collected seeds and sent them back to France, from whence they were introduced as garden plants across Europe.[2]

Native tribes showed early European settlers that mayapples were both a wild perennial fruit and a traditional medicinal plant.[3] Different tribes created poetic names to describe the appearance of the plant as well as the medicinal uses of this distinct-looking perennial. The

A stand of mayapples.

Cherokee tribe used *oo-nee-squa-too-key* (it wears a hat) for the umbrella-like growth, and the Osage tribes called the plants *che-sa-ne-pe-sha* (it pains the bowels), in honor of the toxic, purgative properties of the roots.[4] Although all parts of mayapples except the ripe fruits are poisonous, the roots became a popular pharmacy ingredient in many drugstore medical products by the late 1800s. Extracts of mayapple roots were sold as an ingredient in the popular over-the-counter drug called Carter's Little Liver Pills, which were advertised as being able to "relieve distress from indigestion and too hearty eating, dizziness, nausea, and to regulate the bowels."[5]

Despite all of the outlandish claims made for a toxic root that was advertised as a "miracle drug," by the late 1960s serious medical efforts went into creating semi-synthetic drugs (podophyllotoxin) derived from mayapple roots. These drugs are currently being used extensively and effectively for the treatment of different types of cancer.[6] Wild-harvesting of mayapples for anti-cancer treatments and other medical applications has become a small-scale industry. In 1970, 130 tons of American mayapples were dug from the wild for pharmacological use.[7] Because of this demand, the USDA and several agricultural growers are looking into the commercial production of mayapples for the pharmacological industry.[8]

Medical applications aside, if you have an empty, shaded garden area with dappled light, these plants will provide you with an attractive spring perennial ground cover and a taste of an interesting late-summer fruit.

Growth Difficulty Rating: 1. Mayapples are easy to grow in any environment that gets consistent moisture and has a fair amount of shade.

Taste Profile and Uses: The fruit is a fleshy, lemon-shaped berry about 1½ to 2 inches (3.8–5.1 cm) long. Mayapples are not edible until they soften and start to wrinkle, which is toward the end of August in our gardens. The heat of the summer starts to bring on senescence (plant foliage dying away), and mayapple leaves turn a washed-out, pale yellow-white. This is when the fruit is usually ripe and ready to pick. The pulp of the ripe fruits is delicious and unique, with a tropical flavor that is somewhat like a combination of passionfruit, guava, and pear. The ripe fruits can be eaten out of hand—but discard the seeds and skin, which are slightly toxic. It is best to enjoy the fruits in small quantities because eating large amounts can have a laxative effect on the body. Mayapple jelly is the traditional way to use a large harvest of the fruits. But remember, only ripe fruit is safe to eat; the unripe fruit, seeds, leaves, and roots of this plant are considered highly toxic when eaten.

Plant Description: This herbaceous perennial emerges from the ground in late April to early May, in the form of a 12- to 18-inch (30.5–45.7 cm) stalk encircled by a closed green umbrella. The leaves unfurl into the shape of two dark green, shredded, umbrella-like leaves that grow up to a foot wide. Mayapple leaves are arranged around a light green glabrous (hairless) stalk, which is attached to the underside of the leaf. The tropical-looking leaves are divided into five to nine coarsely toothed lobes per leaf, depending on the maturity of the particular plant. Each umbrella-like set of leaves produces a single flower per season, which can often be hidden by the large growth of the foliage. Plants with a single leaf are immature and will not grow a flower.

As the fruits develop and the energy of the plant goes toward ripening the fruit, the foliage begins to turn yellow until it dies back to the ground. Then mayapples stay dormant through the rest of the season until they begin to grow again the following spring.

Flowers: A solitary, nodding white flower grows in the axil of the leaves where the two petioles (leafstalks) branch out from each other. The 1½-inch (3.8 cm) flower has six to nine waxy white petals, with many stamens that have white filaments and yellow anthers. Up close, the flowers are pleasantly fragrant and can be quite showy, but they are often obscured by the large leaves.

Pollination Requirements: Mayapple flowers do not produce nectar but rely upon infrequent visits by pollen-seeking queen bumblebees and perhaps other species of bees for pollination.[9] Since mayapples are not self-fertile, they need to be cross-pollinated by a separate, genetically different plant that is not growing on the same rhizome root system in order to set seed and to produce fruit. This can be a problem, since many large colonies of mayapples are actually composed of a single long creeping rhizome.

Site and Soil Conditions: This native woodland perennial can be found growing in mixed moist or dry deciduous forests, shaded fields, and riverbanks. Sometimes large patches can be found growing along a forest clearing or in a disturbed area on the side of a road.

Hardiness: Mayapple is hardy to zone 4, or −30°F (−34°C).

Fertilization and Growth: Mayapples will thrive in an environment that reproduces the habitat of their wild growing conditions. This

Mayapple flower.

Unopened flower bud.

Developing fruit.

Ripe mayapple.

would be anywhere from a part-shade to a completely shady location. The perfect conditions would be three to four hours of morning sun to provide them with enough light to encourage the plants to flower, but enough shade to allow them to survive during the hotter parts of the summer. The plants will grow in a wide range of soil types from sandy loam to clay-based soils as long as they are well drained.

Mayapple rhizomes should be planted a few inches (5–10 cm) below the soil level, and the ground enriched with compost or leaf humus to encourage vigorous growth. They should be watered well for the first year after planting but should require no extra effort after the first season as long as they get a moderate amount of moisture from rain throughout the year.

The mayapple plant grows on a long, horizontal root system referred to as a rhizome, and given the right conditions over time, a single mayapple plant can grow into a giant colony that forms an attractive woodland ground cover.

Cultivars: At the time of this publication, there are no cultivars of American mayapple.

Related Species: American mayapple is the only species in this genus that is native to the United States, but there are several other species native to Asia, including the Himalayan mayapple (*P. hexandrum*) and Chinese mayapple (*P. pleianthum*).

Propagation: The easiest way to propagate mayapple is by root division in the fall while the plant is dormant. This means that you will need to mark where the plant was growing since it is an herbaceous perennial and the parts above the soil die to the ground in late summer or after fruiting. When dividing the rhizomes, make sure that each root section has at least one attached bud growing on it. The plants can be dug up in early spring when the stalks first appear and are just poking out of the soil, but if they are beginning to grow more than a few inches (5–10 cm) from the soil, they can experience transplant shock when you replant them.

Pests and Problems: This perennial does not seem to be bothered by insects or diseases. Box turtles, opossums, and raccoons are said to love the fruit, and in our gardens the chipmunks seem to know when it's ripe. You would think that these would be deer-resistant plants because of the toxic chemical compounds in their leaves, but every year in our gardens, an unprotected stand of mayapples gets most of their leaves chewed to the ground.

Maypop

Passiflora incarnata

Apricot Vine, Holy-Trinity Flower, Purple Passionflower, Wild Passion Flower

> *The Maypops of the Southern States are familiar to the children of that region, the fruits, from midsummer to autumn, as large as a hen's egg and somewhat suggesting a lemon but with little nutriment. They are mildly sweet and acid, more* eatable *than edible.*
>
> —Fernald and Kinsey, 1958[1]

There are over five hundred different species of passionflower, the majority native to South America.[2] These beautiful vines are extensively cultivated for both their exquisite flowers that come in a diverse range of colors, as well as for their sweet tropical fruits.

In the late fifteenth century, the Spanish and Italian missionaries who came to the Americas literally "christened" these fruits with the name of *The Flower with the Five Wounds* or the *Holy Trinity Flower* because they thought it had been placed by God among the non-Christian Natives as a teaching tool to help in their conversion.[3] But way before

European missionaries showed up, the Indigenous tribes regularly harvested these plants as food and medicine, and archaeologists have discovered passionflower seeds at prehistoric Native American settlements that are five thousand years old.[4]

All parts of this vine were used for traditional tribal medicine, especially the roots, which were used as a poultice for boils, cuts, earaches, and inflammation. A tea made from the dried leaves was used to treat insomnia—these are still sold as an herbal product in the health food industry today.[5]

In many ways maypop seems more widely appreciated as a medicinal plant than for its delicious tropical-tasting fruit. Maypops (*Passiflora incarnata*) can be found growing throughout the midwestern United States, from Pennsylvania south to Florida and west to Oklahoma and Texas, and it is the only reliably hardy vine in the *Passiflora* genus that produces edible fruits. The actual meaning of its common name of *maypop* seems to depend on which source you are inclined to believe; it might be derived from the fact that the vines "pop" out of the ground in May, or it could stem from the popping sound that the ripe fruits produce when you squeeze or accidentally step on them.

These vines are spectacularly beautiful, and any garden would be improved with a trellis of maypop flowers, even if you never got a single fruit!

Growth Difficulty Rating: 3. Growing maypop is actually easy; in the sixteen years of having this vine, we have never seen a single pest. The real difficulty is getting the plants established, and then protecting them each winter against the coldest temperatures. The plants will eventually require some pruning to keep the vines within the boundaries of most garden settings.

Taste Profile and Uses: The fleshy fruit is technically a berry that is about the size of an egg. Maypops start out with a firm outer shell-like skin in a dull waxy green color, which slowly develops into a larger fruit with a deep greenish yellow color. The fruits are climacteric, meaning that they will continue to ripen after being picked, provided that they have grown long enough on the vines.[6] The true sign of a mature maypop is when the fruit begins to soften and wrinkle on the outside. Ripe fruits become heavy and spongy, and when popped open they will reveal a collection of dark, edible seeds that are surrounded by a series of translucent, creamy white seed coatings called *arils*, which have a deliciously mild, tangy flavor.

The magnificent flowers.

'Alba' flower.

Maypop diversity.

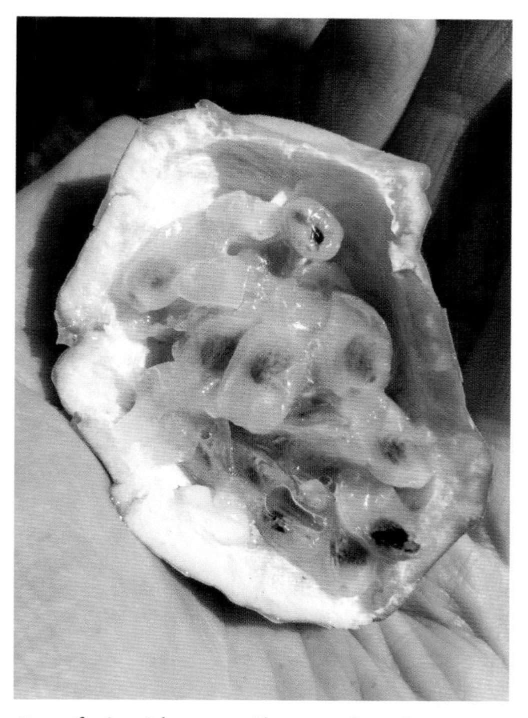

Open fruit with tasty aril-covered seeds.

We like to eat the fruit by cutting it in half and scooping out the pulp with a spoon, but in good years our plants have yielded over a hundred berries, so we have used them to make jellies, sauces, and salsa. Although maypops have many of the great culinary qualities of the tropical passionfruit (*P. edulis*), they lack the acidic citrus-like notes that make the fruit so distinctive. An easy remedy for this deficit is to combine freshly squeezed lemon or orange juice and a sweetener with the maypop pulp when cooking with it.

Plant Description: These are fast-growing perennial vines with handsome, rich green, trilobed leaves that have lightly toothed edges, that are alternately arranged on the stem and range in size from 3 to 6 inches (7.6–15.2 cm) long. The vines produce long tendrils, which are actually modified leaves, and these long, curly, wire-like growths wrap around anything in their path. This allows the vines to grow up vertical structures as they reach toward the light. Mature vines can grow 25 feet (7.6 m) a season and die back to the ground in cold northern climates. In warmer regions the vines may not die back each year and have a tendency to be weedy, so we are grateful that the vines are killed to the ground and start over each season! This is a vine that will require a sturdy support system. Established plants can have deep roots, extending as far as 3 feet (91 cm) into the ground, so consider a planting site carefully because they can be difficult to move.[7]

Flowers: Maypop flowers are andromonoecious, meaning one plant can be either perfect (having both female and male organs) or staminate (male parts), depending on environmental stresses (such as animals browsing on leaves or low light conditions). Stress can make plants produce only male flowers to reduce fruit-set and conserve energy resources.[8] The petals and sepals range in color from white to lavender and are arranged behind the coronas (long, wavy, hair-like filaments) that have alternating bands of dark purple, lavender, and white. Above these are five showy stamens that are flecked with maroon, and are topped by a small, light green ovary with three curved styles and stigmas (the pollen receptacles). The buds of the flowers arise from the leaf axils and begin to bloom in June inside our greenhouse and in July outside it. The blossoms are about 2½ to 3 inches (6.4–7.6 cm) across, and the vines set flowers every day throughout the summer until they are stopped by cool autumn temperatures.

Pollination Requirements: The flowers are pollinated by carpenter bees, and the vines are self-sterile and require a genetically different plant for pollination, so make sure you are buying two separate plants. Old, established vines can send root runners underground 20 feet (6.1 m) or more away from the original plant, and if these shoots are propagated, they will be a clone of the first plant and self-sterile, unable to produce fruit.

Site and Soil Conditions: In the wild, maypops grow like weeds; however, if you want fruit and lots of flowers in a garden setting, we believe that in the northern parts of the United States the vines should be planted only in full sun (from eight to twelve hours,) in the highest-quality garden soil that is well drained and amended with large amounts of compost or leaf humus.

Maypops need a long growing season in full sun in order to ripen, and when that happens we get mature fruit, but cool evening temperatures in autumn can prevent the fruits from fully ripening. This can be helped by draping a clear plastic sheet over the vine to create a mini greenhouse, or picking some of the ripest fruits and allowing them to fully ripen indoors.

Getting Root Systems Established: Most of the plants available in commerce are through mail-order nurseries, which sell small plants with tiny root systems that are not resilient enough to survive cold winters. To grow maypops successfully, the plant's root system needs to be large and thick enough to be able to withstand long freezing temperatures, but once the root systems are large, the vines are more cold-tolerant. Here are two different ways to get them established.

1. Keep a small plant in its pot and grow it outside during the first year, remembering to keep it well watered. Take the potted vine

The almost ripe fruits.

inside for the winter before a hard frost and place it by a sunny window. After two or three years, the roots will get larger and the vine can be planted out in spring.

2. Plant a young vine in early spring right after all danger of frost has passed, so that its roots have a full, long growing season before winter. In the fall bury the plants in 6 inches (15.2 cm) of mulch to protect the root systems, or put plastic buckets over them to protect them. If you do this successfully for two to three years, the roots will get huge and will be less prone to damage in winter.

Hardiness: Maypops are an evergreen vine in zone 8 (10°F/−12°C) or warmer; in zones 6 (−10°F/−23°C) and 7 (0°F/−18°C), the plant behaves like a perennial vine that dies to the ground every year. These plants can been grown in a zone 5 (−20°F/−29°C) environment with winter protection.

Fertilization and Growth: The vines need to be supported on a fence, trellis, or arbor, and the plant's height and spread varies depending on the structure that it climbs on and where it is sited. Maypop has aggressive tendencies, so it is important to site this plant correctly—over time its suckering roots may invade other plant beds, and can completely grow over other garden plants like a pushy, belligerent weed. Fertilize the vines every spring with compost or leaf mulch, but do not use manure, which will grow lush foliage at the expense of the flowers.

Cultivars: There are several cultivars for *P. incarnata*, and our maypop vines produced much more fruit when we planted a white maypop vine with them, which promoted pollen diversity.

'Blue Pop': A new cultivar of maypop with flowers that have a blue-purple cast.

'Pink Pop': A new cultivar with more pronounced lavender-colored flowers.

White Maypop 'Alba': A beautiful pure white form of the native maypop.

Related Species: The yellow passionflower (*P. lutea*) is another native vine in this genus and has 1-inch (2.5 cm) wide pale-yellow flowers.

Propagation: Maypop vines can be started from seed, or shoots from around the base of the vine can be dug up and grown in pots until they have produced large root systems before planting out.

Pests and Problems: We have not had any pests or disease issues with our vines.

Medlar

Mespilus germanica

Showy Mespilus

His wit is like a medlar; it is never ripe until it's rotten.

—Samuel Johnson, 1791[1]

Medlar is an ancient fruit brought in the Middle Ages from Turkey to Europe, where it spread as an exotic dessert fruit. Medlar became an extremely fashionable cultural "name drop" for authors and artists in Europe. This fact is borne out by the hundreds of references to medlar that are found in the written works of antiquity.[2] The depiction of medlars in European art masterpieces also demonstrates the popularity of the fruit. They were used as decorative elements in works of calligraphy from the 1500s; they show up along the borders of illuminated religious manuscripts and as part of the ornate displays of sumptuous fruit arrangements in Dutch still-life paintings.[3]

This once popular fruit is now mostly unheard of, but it can still be found growing in older European monasteries and gardens. Medlar rarely made it across the Atlantic to the United States until recently, and there may be several reasons for this. Perhaps the old medieval

Beautiful flowers in spring.

Developing fruit.

'Royal' fruits ripening.

Ornamental winter fruit.

name *openarse* for the fruit's supposed likeness to a horse's rear end is not very supermarket-friendly. The lack of interest in the fruit may also stem from the method of ripening it. Before medlars are edible, they must go through a ripening period called bletting—the process of the fruit softening beyond the point of normal ripeness. While this is not actually rotting, it is very close to it. In the fall after the hard fruits are softened by either frost or by several weeks of careful storage, both the consistency and the flavor become similar to applesauce. Despite these issues, medlar has gotten a little more attention the last fifteen years or so, as it's started to be sold in nursery catalogs and appear in garden writings.[4] A few years ago we even saw it artistically displayed in a high-end food market in Manhattan called Dean & Deluca, for a whopping $19.95 a pound.

All of these characteristics might make some people think that medlar should best be described as an acquired taste, but we feel that it is a worthwhile small fruiting tree for the home gardener. The plants are very ornamental because they feature attractive large, white blossoms and abundant displays of red-russet-colored fruit. Best of all, medlars often persist on bare tree branches throughout much of winter, adding another season of ornamental interest, especially if they get decorated with snow and icicles. It is wonderful walking through a winter landscape and being able to stuff our coats with pocketfuls of frozen fruit when nothing else seems alive.

Growth Difficulty Rating: 1. Medlar trees are easy to grow, with few to no pests.

Taste Profile and Uses: The 1½- to 2-inch (3.8–5.1 cm) fruit is a pome that looks like a large brown rose hip with an open calyx that is lined by sepals (modified leaves) on the bottom of the fruit. The skin of the unripe fruit is a pale tan-yellow overlaid with fine bumpy scales, and as the medlars mature, they change into a dark russet color in the fall. The insides of the fruit are hard and white while the fruit is young, but when ripe the pulp develops into an ocher color. Some people may find eating medlar fruit a bit crude: Although you can use a spoon, we think the fruit is tastiest eaten standing in front of the tree, squeezing out the sweet pulp from the backside while spitting out the inedible seeds and skin.

One negative aspect is the inconsistent nature of the fruit. Each year our tree produces lots of medlars that are perfectly formed with smooth-textured pulp that is rich and delicious, but there are always some fruits that are less well formed, with a grainy, unpalatable taste. When the fruit is at its best, it has an applesauce-like consistency and taste, with hints of cinnamon and raspberry. In Europe the fruit is traditionally used to make medlar "cheese" with a similar texture to quince paste. Medlar jam is also popular in Europe, and the fruit could be combined in sauces with other fruits.

Plant Description: A large, multi-trunked shrub or small tree whose trunk, branches, and limbs often grow into an interesting twisted and crooked form, which could be mistaken for an apple tree. The plant has handsome, tropical-looking, deep green foliage whose 3- to 5 inch (7.6–12.7 cm) leaves are toothed, and oblong to lanceolate in shape. Though there are some ancient medlar trees in Europe that are 20 feet (6.1 m) tall, most will reach between 8 and 10 feet (2.4–3.1 m), so it is a good choice for a small yard or modest-sized

The shrub-like habit of medlar.

garden space. Once the fruits ripen in the late fall, they can be left on the tree and picked throughout winter. This makes them a distinct and attractive addition to a winter garden landscape.

Flowers: The beautiful white flowers have five petals with a ruffled appearance, and are 1½ to 2 inches (3.8–5.1 cm) in diameter. Directly behind the flower petals, the long, triangular woolly sepals are pointed and extend past the blossoms, which collectively make up the calyx (collection of sepals) surrounding the opening on the front of the medlar. The camellia-like flowers open from late May to June in our gardens and are one of the most attractive aspects of this tree.

Pollination Requirements: Medlars are fully self-fertile and require only one tree in order to produce fruit.

Site and Soil Conditions: Medlar trees like full sun but seem to do well in a part-shade location, too. We have a nineteen-year-old tree sited in four to six hours of morning sun that consistently bears fruit every season. Medlars seem to grow in most soil conditions as long as they are not permanently wet.

Hardiness: The trees are hardy to zone 4, or to about −30°F (−34°C). Medlar trees require some freezing temperatures in order to bear fruit.

Fertilization and Growth: Medlars are slow-growing trees, but they will be more vigorous and productive if provided with an annual application of compost or leaf humus as a mulch.

Cultivars: There are roughly a dozen different cultivars of medlar available through mail-order catalogs and nurseries.[5] Below are a few of the popular varieties.

'Breda Giant': An old, traditional cultivar from Holland that consistently produces abundant crops of large fruit.

'Macrocarpa': A traditional cultivar from Europe, famous for producing good-quality fruits that are among the largest medlars—2½ inches (6.4 cm) in diameter.

'Royale': A productive cultivar of medlar that produces high-quality fruit.

'Sultan': A heavy-cropping, large-fruited cultivar from the Netherlands introduced into the United States by the National Clonal Germplasm Repository in Corvallis, Oregon.

Related Species: Medlar is in the rose family and is regarded as a very close relative to hawthorns (*Crataegus* spp.). Indeed, hawthorn fruit has the same general shape as a medlar, including the calyx on the front of the fruit, but the medlar is larger. The two genera *Mespilus* and *Crataegus* are so closely related that some hawthorn species can be hybridized with medlar.[6]

Sterns medlar (*Mespilus canescens,* or *Crataegus* × *canescens*): This is one of the rarest wild plants in America. It was discovered in Slovak (Prairie County), Arkansas, in 1969. There are about twenty-five known bushes of this extremely rare medlar plant, which is one of the great scientific curiosities of American flora. It is unclear whether these plants are spontaneously generated crosses between two hawthorn species, or were brought to Arkansas by eastern European immigrants in the 1800s and developed into a separate species of medlar.[7]

Propagation: Medlar is traditionally grafted. Our older medlar tree is grafted onto a quince rootstock that produces shoots near the bottom of the tree. If you know that your medlar tree is grafted on quince rootstock, you can remove the root suckers from the quince with a small amount of root attached to them and pot them up in soil for a few years. The net result of this can be extra quince rootstock for grafting more medlar trees. Just make sure that you are not cutting into any large roots, which could injure the tree. The other rootstocks that are suitable for grafting are medlar and pear, as well as its close relative hawthorn.

Medlar can also be grown from seeds, but medlar seedlings will usually not end up producing high-quality fruiting trees that are as reliable as a grafted cultivar. Taking softwood cuttings of medlar in early summer can produce trees, but this method tends to be less successful than grafting.[8]

Pests and Problems: Medlars are disease-resistant and virtually pest-free. Over the past few years we have noticed that rust (see the 'Pear' chapter) is appearing on some of our fruits. Fallen fruit may be chewed on by animals, but we have found that during the winter the fruit stays on our trees and generally is not bothered by birds.

Mulberry

Morus rubra, M. alba

Red Mulberry, White Mulberry

> *. . . if all the highways in country towns were ornamented with a row of mulberry trees, on each side, half a rod apart, each mile would contain 1380 trees, the income of which, after seven years, would probably pay for repairing all the highways and the expenses of the public schools. . . .*
>
> —Mr. Cobb, 1826[1]

There were many high hopes for the silkworm industry in America that did not pan out, but the human race has had a long history with mulberry trees that goes back to the writings of the Roman poets Ovid and Virgil, and to the introduction of mulberries as silkworm food in Constantinople in 527–565 CE.[2] The white mulberry (*Morus alba*) is native to eastern and central China, where it has been widely cultivated for hundreds of years as an agriculturally important tree because the leaves are used to feed silkworms.[3] White mulberry was introduced into America for silkworm culture in early colonial times; one of the first legislative acts passed in Virginia during that period required every landowner to plant six mulberry trees each year over a seven-year period, and forbid the destruction of mulberry trees in the clearing of forests, in order to promote the young silk industry.[4] Although the silk industry proved to be unprofitable and was abandoned, the white mulberry continued to flourish, and ended up becoming an invasive pest species in almost all fifty states.[5]

The taste of the fruit from this species is what most people associate with the mulberry, which is sweet and insipid. Or they associate the taste of mulberry with the "watery blackberry" flavor of the native red mulberry (*M. rubra*), with the quality of the fruits ranging from good to poor. But if they were to taste the flavor of a cultivated variety of mulberry, it would be a completely different experience. We feel that mulberries are among the best fruits in the world. The flavor of a good variety is complex and, once you taste the fruit, may be difficult to describe. The best varieties elude easy classification and are a strong blend of sweet and tart flavors, which can approach blackberry, raspberry, tangerine, and pomegranate all mixed together!

If you have a planting area that could fit a large-sized tree, then a mulberry is well worth the effort, providing large crops of summer fruit.

Growth Difficulty Rating: 1. Mulberry trees are easy to grow with few pest problems.

Taste Profile and Uses: Mulberry trees produce fruit on the current season's new growth and the spurs (short fruiting stems) on the older branches of the tree, with fruit maturing from late June to July; some "everbearing" cultivars produce a second flush of fruit into August. Mulberries are not technically considered berries but a fruit cluster composed of closely packed fleshy drupes.

Our favorite way to enjoy the fruit is to stand under the tree and pick them off the branches, but if you are lucky enough to have a big harvest, a worthwhile but messy endeavor is to spread a sheet under the tree and then shake the limbs gently to release the ripest fruits.

Mulberries can't be sold in food markets because of their short shelf life; even with refrigeration the fruit will last only a few days. The best remedy for this problem is to make mulberry jam, wine, or even pies. The fruits dehydrate well and can be used in granola, or the fresh berries can be frozen for long-term storage.

The leaves make a pleasant tea, and fall-harvested leaves have been used in traditional Chinese medicine.[6]

Types of Mulberry: There are multiple species of mulberry and some confusion about their correct identification, but in general three primary species are cultivated for fruit. The Latin name of each fruit does not signify its actual fruit color, but is a species designation: the black mulberry (*M. nigra*), the red mulberry (*M. rubra*), and the white mulberry (*M. alba*). Because black mulberry cannot be reliably grown in the northern parts of the United States, we will profile only the other two species.

Red Mulberry (*M. rubra*): The 1- to 1½-inch (2.5–3.8 cm) long fruit is deep red to a purple-black color with a watery blackberry flavor that won't knock your socks off.

White Mulberry (*M. alba*): The fruit is highly variable, and there are cultivars with white, pink-violet, purple-black, and black-colored fruit. The fruit of the straight species is generally sweet, bland, or mediocre with no unique characteristics, but there are dozens of great cultivars of white mulberry that produce excellent-quality fruit. The size of the fruit is also variable, from about ½ inch to 1½ inches (1.3–3.8 cm) long, depending on the particular tree and cultivar.

Plant Description: Mulberry trees are either dioecious (separate male and female trees) or monoecious (trees with both male and female flowers), but sometimes the trees will change from one sex to the other.[7] The trees are characterized by a short trunk with limbs that divide into numerous branches and gives their outline a dense, rounded form. The trees feature medium- to large-sized shiny, green leaves that are variously lobed, with certain trees producing mitten-shaped leaves while other trees have small, scalloped lobes. In the wild, mulberry trees grow fast, and both the red and white mulberry trees can reach 50 to 70 feet (15.2–21.3 m) in height, but grafted cultivars are often much shorter, reaching 20 to 30 feet (6.1–9.1 m).

Flowers: Mulberry flowers are tiny and often go unnoticed. Both the male and female flowers are held on short, nondescript catkins. Male flowers carry the pollen on densely arranged flowered spikes (catkins) about an inch (2.5 cm) long, hanging from short peduncles (stems) in the axils of the leaves. There are often a few female flowers mixed in with male flowers, but the female catkins are generally twice as long as the male flowers and are borne out in the leaf axils of the very first sets of leaves.[8] The female flower looks like a small green mulberry; each has little tentacles arranged on top to collect the pollen.

Young 'Illinois Everbearing' tree.

The thread-like flowers.

'Tartarica' fruit.

Ripe 'Illnois Everbearing' fruit.

Pollination Requirements: Mulberries are self-fertile and are pollinated by wind, but there are some cultivars that are parthenocarpic (develop fruit without pollination), which will result in seedless fruit or seeds not being viable.[9]

Site and Soil Conditions: Red and white mulberries are considered to be weed trees because they can grow in almost any conditions except soil that is constantly wet. Red mulberry is able to grow in part shade to full sun, but white mulberry does not do well except in full sunlight. For the most productive harvests and the least amount of disease problems, we recommend a planting area in full sun with soil that has been amended with compost or leaf humus. Mulberry trees are often used as street trees in cities because they can tolerate pollution, and established trees can tolerate drought. Choose an open growing area away from the house because the fruits will eventually cover the ground with messy purple-black pulp that can be tracked indoors on shoes.

Mulberry Root Systems: Mulberry trees have extensive horizontal root systems. When they are grown by nurseries, their roots often circle the insides of the pots. When planting a mulberry tree, it is very important to carefully tease out the pot-bound roots and make sure that they are not encircling the tree trunk, but point away from the trunk and into the soil. The best time to do this is in the spring before the dormant buds on the trees have opened; this will help you gently tease apart tangled root systems so that the trees will not experience transplant shock. We have purchased several pot-bound mulberries with twisted root systems that were completely strangling the trunk of the trees. If we'd left them in place, an otherwise healthy tree might have actually been killed by its own root system.

Hardiness: The white and red mulberry and their various cultivars are rated at zone 5, or −20°F (−29°C), but some cultivars of these species are hardy to zone 4, or −30°F (−34°C).

Fertilization and Growth: Mulberry trees grow fast and should be provided with a mulch of compost or leaf humus every year to provide nutrients for the soil while protecting the roots from drying out. Mulberry trees can bleed excessively when they are pruned, so the only time to do major pruning is during the winter when the trees are fully dormant.

Cultivars: We recommend planting only the cultivated varieties that have been selected for high-quality fruit, because mulberries produced on the straight species are inferior. There are currently over a dozen grafted varieties of mulberry derived from *M. rubra, M. alba,* or the hybrids between the two species. Below we have listed a few of the high-quality cultivars.

'Illinois Everbearing' (*M. alba* × *M.rubra*): A cross between the red and white mulberry that produces superb-quality fruit with a perfect balance of sweet and tart. The trees have done well in many parts of the United States. They are called everbearing because they produce two batches of fruit.

'Oscar' (*M. alba* × *M. rubra*): Another cross between white and red mulberry producing delicious fruit that ripens early in the season.

'Weeping Mulberry' (*M. alba*): A small 15-foot (4.6 m) tree with attractive weeping branches and sweet fruit.

Related Species: Mulberries are related to the common fig tree (*Ficus carica*) as well as jackfruit (*Artocarpus heterophyllus*) and che (*Maclura tricuspidata*).

Propagation: Grafting mulberry scions (young hardwood cuttings) on white or red mulberry rootstocks is the traditional way to propagate new trees for cultivated varieties, although softwood cuttings taken in summer will occassionally form roots.

Pests and Problems: Mulberries are generally free of pests and diseases. The popcorn disease of mulberry is caused by a fungus (*Ciboria carunculoides*), which affects trees in the southern parts of the United States, in zone 7 (0°F/−18°C) or hotter, but this is currently not a problem for trees planted in growing zones 4 to 6 (−30 to −10°F/−34 to −23°C).[10]

Mulberries are attractive to birds, so you may be competing with them to get the fruit.

Nanking Cherry

Prunus tomentosa

Chinese Bush Cherry, Downy Cherry, Manchu Cherry

The cherries are a half-inch in diameter, bright currant-red, covered with inconspicuous hairs, are pleasantly acid, very juicy, and a great addition to cultivated cherries. P. tomentosa *seems a promising plant for domestication and of particular merit for small gardens.*

—U. P. Hedrick, 1922[1]

The Nanking cherry is native to Korea, Tibet, and the northern Himalayas, as well as northern and western China. These cherry bushes can be found growing wild in mountainous areas that experience brutal winter temperatures as low as −30°F (−34°C).[2] It was cultivated throughout China and Japan as an ornamental shrub because of its bountiful batches of cheerful, early, pinkish white flowers and for its decorative, small cherry fruits. These attractive attributes, combined with the plant's ability to do well in cold regions, are the reason these plants were brought into North America as useful garden shrubs in the early twentieth century.[3] This species' initial introduction into the United States was through the Arnold Arboretum in 1882, from seeds gathered in Peking, China. From 1903 through 1953 the USDA Plant Introduction's missions continued to collect and introduce seeds from multiple cultivars of the Nanking cherry.

One of the people responsible for collecting improved cultivars was the great plant explorer Frank Meyer, who sent hundreds of Nanking cherry seeds back to America because he felt that this low-maintenance fruiting plant had tremendous commercial potential.[4] As people started to migrate west in the early part of the twentieth century, plant breeders were interested in developing cold-hardy fruits that could withstand the conditions of the northern plains, which experience frigid winter temperatures and periodic prolonged droughts. By the 1930s Nanking cherries were being promoted as "the crop of the future" in the Midwest.[5] By the late 1940s, however, none of these efforts materialized and the shrubs were relegated to a few arboretums and fruit orchards. This was probably because of the fruit's small size compared with the popularity of commercial sweet cherries (*Prunus avium*), or maybe due to the fact that Nanking cherries deteriorate quickly when ripe and would not last on market shelves.[6]

Whatever the case, it's unfortunate that the Nanking cherry has never been more widely planted. When the small fruits sparkle like little red jewels and completely cover the branches of the bushes in midsummer, they are more than just decorative, they are delicious to eat.

Growth Difficulty Rating: 1. It is easy to grow Nanking cherries. Although they are not immune to all of the major pest problems that plague the *Prunus* genus, they are much less affected by them.

Taste Profile and Uses: The fruits are considered drupes and are less than ½ inch (1.3 cm) across with an almost circular shape. Their skins are an attractive, deep scarlet-red color that is covered in very fine hairs. In our gardens the fruits begin to mature in late June to early July.

The overall flavor of a ripe fruit is most similar to a pie cherry (*P. cerasus*) and contains a good amount of juice and a pleasant blend of sweetness with a mellow tart finish. The cherries need to mature on the bush, and ripe fruit will often fall from their pedicel (fruit stalk) when you tickle them with your fingers. One practical method of gathering a large harvest is to lay a sheet underneath the plant and gently shake the limbs to dislodge ripe fruits. Although we like to eat the small fruits out of hand, a mature bush can produce

White Nanking cherry blossoms.

The elegant small flowers.

White Nanking cherry fruits.

The jewel-like fruit and handsome leaves.

copious amounts, great for jelly, pies, and beverages. There is a pit inside each Nanking cherry, and removing these from the fruit for cooking can be an onerous task. The most convenient way to do this is by using a kitchen ricer to separate the seed from the pulp.

Plant Description: *Prunus tomentosa* is a vigorous, densely stemmed deciduous shrub that ultimately reaches between 6 and 12 feet (1.8–3.7 m) tall or more, with a tendency to grow horizontally. The attractive 2- to 3-inch (5.1–7.6 cm) long leaves are obovate in shape with a deeply serrated texture and toothed leaf margins; they end in acute pointed tips and are alternately arranged on the branches. The leaves are a rich green color and the undersides as well as the young branches are pubescent (hairy), which is why this shrub's Latin name is *tomentosa,* which translates into "covered in hairs." Even the fruits are lightly fuzzy, but you really need a magnifying lens to see them. The shrubs grow smooth grayish-brown limbs that have a dense branching habit, which makes these plants suitable for a windbreak, wildlife habitat, or a good privacy hedge between houses. The shrub's trunk has exfoliating, metallic-brown bark that peels away in strips, making it very handsome to look at in the winter.

Flowers: In early spring the small pointed buds become more prominent until they open up in masses of ¾- to 1-inch (1.9–2.5 cm) flowers, borne singly, in pairs, or in small clusters on short

Magnificent early-spring flower display.

pedicels. The small, delicate pinkish-white flowers develop on one-year-old wood as well as older branches, just as the leaf buds are cracking open. The branches can be completely covered with blossoms due to the abundant numbers of spring buds and their close arrangement on the stems.

In fact, when visitors to our gardens see Nanking cherries in flower they often want to use them in their gardens as ornamental plants and do not realize that they are fruiting shrubs.

Nanking cherries are precocious plants that often produce small sprays of flowers and fruit after they have only reached a few feet in height. Occasionally the early-spring flowers get blasted by late-spring frosts and we still manage to get some fruit.

Pollination Requirements: Nanking cherries are self-sterile and will require another bush for cross-pollination in order to set fruit.

Site and Soil Conditions: Like all plants in the *Prunus* genus, the Nanking cherry needs a full sun planting site for good fruiting. It prefers rich, slightly acidic, well-drained soil, but established plants seem to be highly adaptable to many types of conditions except for wet areas. The shrubs will also tolerate drought after they have been established for a few years.[7] Nanking cherry is traditionally planted as a windbreak in China, where the plants survive heavy amounts of snow as well as long hot summers.[8]

Hardiness: Nanking cherries are the hardiest of all the edible species of cherry and can tolerate temperatures down to zone 2, or −50°F (−46°C), but the flower buds are often damaged by late frosts in growing environments colder than zone 4 (−30°F/−34°C).[9]

Fertilization and Growth: Unlike a lot of fruiting bushes, the Nanking cherry tends to exhibit fast growth rates in its early growing years. As the shrub matures the branches are produced in great numbers and become crowded; they will benefit from an occasional pruning. Removing older branches as well as dead ones will open up more room for productive younger, fruiting limbs and create more air circulation to help the shrubs be more productive. This will discourage foliar diseases by increasing airflow and sunlight to the lower parts of the plant.

Like all members of the *Prunus* genus, Nanking cherry would benefit from a yearly application of compost in spring. If you want these plants kept at a lower height for convenient fruit picking, they are easy to prune and can be kept at a manageable size.

Cultivars: When Nanking cherries were introduced in the Midwest, there were several improved cultivars developed, but unfortunately these have been lost in horticultural history, and now the only cultivar that is widely available in commerce is the white-fruited form.[10]

'Leucocarpa' (white Nanking cherry): A white-flowered form that produces pale yellow fruit that is sweeter than the red ones, with a flavor similar to the 'Royal Ann' yellow cherries.

Related Species: Both the sweet cherry (*P. avium*) and the sour cherry (*P. cerasus*) are closely related to the Nanking cherry.

Propagation: Plants are generally grown from seed, which will result in shrubs that show some variability in fruit quality, crop yield, and winter hardiness. The seeds need to be winter-stratified (exposed to chilling temperatures). Seedlings can start to flower within three to four years.

An easier way of creating more plants is by layering, which is accomplished in spring by bending some of the outer branches down to the ground and shoveling a small hill of soil over the middle portion of each branch so that only the top part sticks out of the soil. After one or two growing years, the part of the branch under the soil will develop a small root system; you can cut below the new roots to separate the layered branch from the mother plant. Pot this up for a year until the roots grow larger and then plant it in a different garden location.

Pests and Problems: Over time members of the *Prunus* genus tend to be afflicted with troublesome diseases such as black knot (*Apiosporina morbosa*) and brown rot (*Monilinia fructicola*). Nanking cherries in many areas of the United States have been mostly free of these fungus troubles, but they are not completely immune from them.[11]

After years of having no problems with Nanking cherry, several springs with heavy rainfall have produced a new problem for us, where the leaves on the branches get discolored, dry up, and fall off the plant, and at the time of this publication we are still assessing how to solve this. For now we have pruned back the shrubs, removing the diseased parts as well as a lot of the older woody growth, to create more air circulation, which helped the following year.

Every year our bushes are used as a habitat by one or two families of nesting catbirds, which are well hidden in the thick, twiggy canopy of the leaves. By the time the fruit ripens in our gardens, the dense foliage covers over some of the fruit and makes it somewhat difficult for birds to completely strip the bushes of fruit.

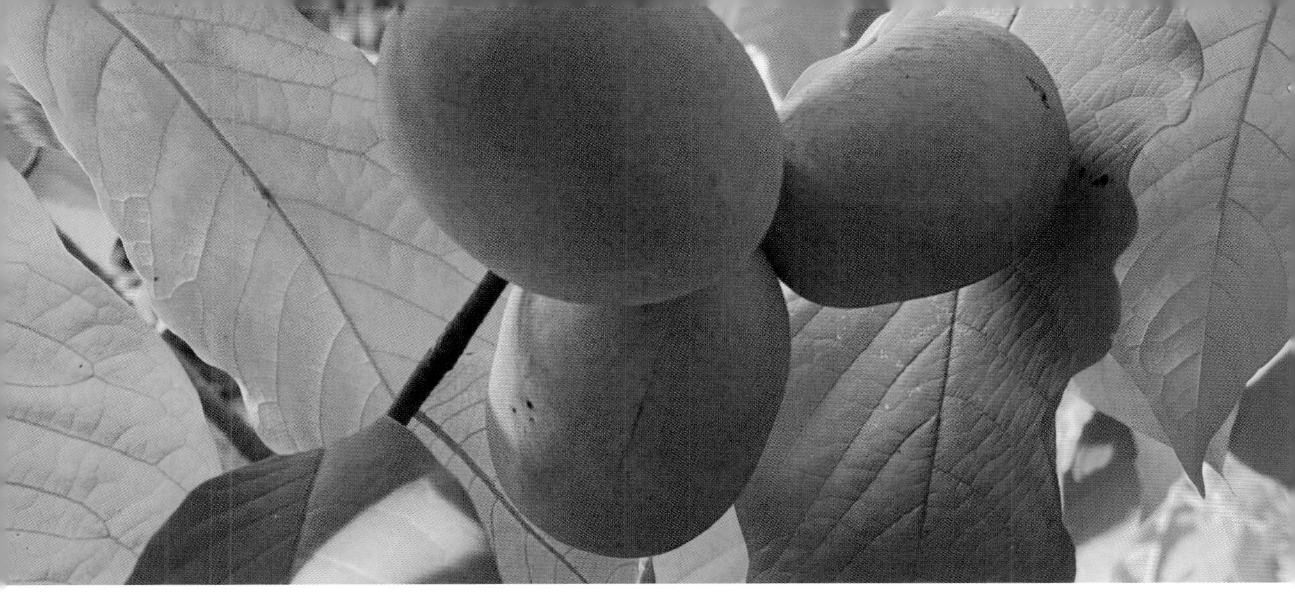

Pawpaw

Asimina triloba

Custard Apple, Fetidshrub, Hillbilly Mango, Indiana Banana, Poor Man's Banana,

> *In its range a characteristic part of American country life, the pawpaw, for all its exotic kinship, seems an intensely native tree, above all in the frosty autumn, when the leaves droop withering on the stem and the great plashy fruits hang preposterously heavy on the twigs.*
>
> —Donald Culross Peattie, 1950[1]

This exotic-looking tree with huge leaves and tropical-tasting fruit doesn't seem like it is native to the United States. Of all the two thousand species of plants in the tropical Annonaceae family, pawpaws are the only genus that managed to migrate out of the tropics and find a home in the cold temperate forests of the US. The geographic range of the pawpaw goes from Ontario in Canada, down to northern Florida, and as far west as the eastern part of Nebraska, and unlike its tropical South American relatives, pawpaw trees actually

The tropical-looking leaves.

Blossom ready for pollination.

A heavy cluster of fruit almost ready to eat.

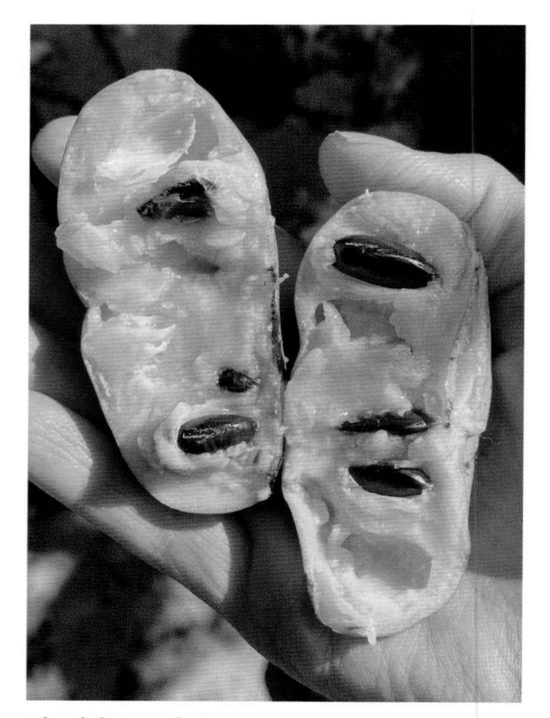

The delicious fruit.

require a minimum of four hundred hours of temperatures below freezing or they will not bear fruit.[2]

These underutilized fruiting trees have a long history on our continent, and current archaeological evidence suggests that pawpaws grew in North America as early as fifty-six million years ago. The fruits were feasted upon by extinct giant mammals, which acted as dispersal agents that geographically spread the seeds via their digestive tracts.[3] Pawpaw seeds have been discovered among the fossils in the Meadowcroft Rockshelter and Historic Village in Avella, Pennsylvania, which is the earliest known site of human habitation in North America.[4]

Pawpaws were loved by George Washington, who planted the trees on his property in 1785; Thomas Jefferson's garden records note that he planted them at Monticello; and there are accounts in Lewis and Clark's diaries that record how they survived on pawpaws alone for three days.[5] In 1900 David Fairchild, the great USDA plant explorer, wondered why America had neglected the one native fruiting tree with the largest fruits?[6] Since the mid-twentieth century, there have been many horticulturalists that have worked on breeding larger pawpaws; most recently, R. Neil Peterson has made major strides in developing modern pawpaw cultivars with giant-sized fruit and improved flavor.

Pawpaw trees take a long time to begin fruiting, but we think the tropical-flavored fruit is well worth the wait.

Growth Difficulty Rating: 2. Establishing pawpaw trees takes some work and patience, but after that the trees are easy to grow with few major problems.

Taste Profile and Uses: The pawpaw is technically a berry that grows into an oblong shape with a light green skin; it is produced either singly or in a cluster. Often compared to the banana in its shape, the pawpaw is actually more cylindrical, but shows some variation in shape and size, ranging from 3 to 6 inches (7.6–15.2 cm) long. By the early part of the fall, the fruits turn a yellow-green color and develop dark brown-black markings as they mature and begin to soften. Inside the skin the fruit's pulp is yellow to orange-yellow with two rows of large, shiny, oblong brown seeds embedded in the pulp. The highly aromatic pulp has a creamy texture, and the flavor is a rich mix of banana and mango with a dash of vanilla, and is delicious!

The fruits can be cut in half and eaten out of hand by squeezing the pulp out, or you can use a spoon if you want to be more genteel. We have also found freezing the fruit to be a wonderful way to preserve large crops; eat them just after they have defrosted a bit so that the pulp is akin to a wonderful custard ice cream. There are many ways to use excess fruit, including to make ice cream, cakes, beer, and wine, but the treat of defrosting a few pawpaws in the middle of winter as a small memory of the previous season's harvest is probably our favorite way to enjoy them.

Plant Description: This deciduous tree typically grows 15 to 25 feet (4.6–7.6 m) tall under cultivation, though larger trees can be found in the wild. The elliptical-shaped leaves start to emerge after the flowers are fully open, and grow 5 to 12 inches (12.7–30.5 cm) long. The foliage turns gold in the fall and the leaves droop down on the branches, which gives the trees a tropical look.

Flowers: The blossoms are formed as small, furry, circular buds on the previous year's wood, and are produced from the leaf axils on pedicels (small stalks) covered with short, rusty-colored hairs. When the buds crack open, the flowers start out green and transition to pink until they are fully unfurled as a strange dark wine color, except for the greenish brown sepals. The unusual triangular flowers are composed of six petals, arranged in two tiers of three that surround three to seven ovaries, which will develop into fruit if they are pollinated.[7] The flowers face the ground and are 1½ inches (3.8 cm) wide, with a fragrance that is often referred to as yeasty or fetid. Their rotten bouquet is meant to entice flies for pollination. The flowers are perfect, having both male and female parts, and they open up at different times during their life cycle.

Pollination Requirements: Pawpaw trees are self-sterile and require two different trees in order to pollinate the flowers and produce fruit. There is a widely held belief that pawpaw trees are only pollinated by carrion flies and beetles, which are attracted by the smell and the color of the flowers, which are thought to resemble rancid meat. But we have noticed many different species of flies and beetles pollinating our trees. However, cool spring temperatures can inhibit insects from flying around when the pollen is ripe, so we always go through the effort of hand-pollinating some of the flowers using a small paintbrush just in case the flies are too lazy to do their job.

Site and Soil Conditions: In the wild these trees grow well in alluvial soil along creeks, streams, and rivers as understory trees. Pawpaws are fairly adaptable and not very particular about soil but would prefer soil rich in organic matter as long as they have good drainage. The most vexing part of growing these trees is that for the first four to six years of life, seedling trees require a partly shaded growing area, but after the trees reach about 4 feet (1.2 m) in height, they need full sunlight for the best fruiting.

Hardiness: Pawpaw trees are found as far north as the bottom part of Ontario, where temperatures can drop to −20°F (−29°C), and are rated to zone 5. Since the fruit must ripen on the trees, the colder parts of zone 5 may periodically experience early frost that damages the fruits before they have fully ripened. So although these trees can survive in this zone, the crops may be occasionally lost. If you are in the coldest parts of zone 5, the best option may be to choose cultivars that have a reputation for early ripening such as 'Allegheny', 'Summer Delight', 'Pennsylvania Golden', 'Mango', and 'NC-1'.

Fertilization and Growth: The trees are slow growing and commonly take as long as ten years to start bearing flowers and fruit. Grafted varieties can produce fruit much sooner and may cut that time in half depending upon the size of the plant. Once established, pawpaw trees have the tendency to send up root suckers and form into large colonies, but when this happens the trees often stop producing flowers, sending their energy towards the roots systems to produce new trees. So these suckering shoots should be cut away near the ground. We spoil our trees by keeping them well weeded, and amend them with compost and leaf humus in the spring and again in the late fall.

In their early years the trees spend most of their energy developing an extensive taproot underground; while, aboveground, the tree displays minimal growth, which means that for several years it looks as if it's not growing very much. Because of their deep taproots, even young seedling trees can be killed by transplanting, so make sure the planting site that you select is the permanent place for the tree. You can buy large trees that are 4 or 5 feet (1.2–1.5 m) tall and plant them in full sunlight; or plant smaller trees in full sunlight with a shade cloth shelter near them to reduce sun exposure; or you can keep repotting small plants every year for many seasons before planting them in the ground. This is a long time to keep potted plants, and it is not easy to nurture seedlings for many years, particularly because of their lengthy taproots.

Pawpaws require a substantial investment of time and grow at a slow rate, which is not in many people's time frame.

Cultivars: In the last ten years, pawpaws have gotten more publicity in the culinary world and in the nursery trade, with fruit breeders introducing many new types. There are now over forty named varieties.[8]

'**Mango**': A cultivar from Georgia that is considered a robust grower, producing high-quality fruits that are an orange-yellow color.
'**Shenandoah**': One of the largest and most flavorful pawpaws that was bred and patented by R. Neil Peterson, producing fruits that often weigh 1 pound (0.5 kg).
'**Sunflower**': A reliable variety with high-quality fruit that won first prize at the 2010 Ohio Pawpaw Festival. Some sources list this cultivar as being self-fertile.[9]

Related Species: The relatives of this plant are mostly tropical, but there are two more American pawpaw species: the dwarf pawpaw (*A. parviflora*) and flag pawpaw (*A. obovata*), both of which are very rare and not reliably hardy in northern parts of the United States.

Propagation: Pawpaws can easily be grown from seeds. The best results will come from fresh seeds that are scarified: The seed coat is gently scraped with a knife to allow water to permeate into the interior of the seed and for its roots to escape. Then these are planted in soil and cold-stratified for sixty days. When a seed opens up, the pawpaw root system can spend a month or more sending out a taproot before the actual plant pushes up through the soil in the pot.

Grafting scions of cultivars onto seedling pawpaw rootstock in spring is also a traditional way to produce trees that are exact replicas of specific varieties.

Pests and Problems: Pawpaw trees are generally disease- and pest-free, although the native pawpaw peduncle borer moth (*Talponia plummeriana*) can be a problem. And when the fruit turns ripe, there may be some competition from hungry raccoons, opossums, and squirrels.

Pear

Pyrus communis

Common Pear, European Pear

> *The origin of the cultivated pear is so completely hidden in prehistoric darkness that it can never be known precisely from what wild pear it came. The historian must content himself with recording what the pear was when written records began . . .*
>
> —U. P. HEDRICK, 1921[1]

The pear genus (*Pyrus*) contains about twenty-five species that are scattered across Europe, northern Asia, and northern Africa. The European pear (*P. communis*) is one of the most widely cultivated trees in that genus and although the trees are found in the forests of southern Europe, and in Asia, the exact geographic origins of this tree are still not definitive because the trees found in the wild may have been cultivated and introduced in those locations.[2] Wherever the actual ancestral location of the common pear really is, so many countries have adopted this fruit that it is now almost global. Since the cultivation of the pear goes back to ancient European culture, it should come as no surprise that it has been hybridized and improved upon since 1000 BCE, and is glorified in classical literature by such writers as Homer, who described the pear as a "gift of the gods."[3]

The European pear has gone through a constant series of revitalizations by the pomologists of various European cultures. France, in particular, has had an infatuation with pears that goes back to 1628, when Le Lectier published a catalog of fruits that included at least 254 pear varieties.[4] Le Lectier was an amateur pomologist whose publication created a country-wide fad for the taste of pears in France.

In the 1800s another amateur pomologist in Belgium named Jean-Baptiste Van Mons also fell in love with pears and cultivated four hundred different varieties.[5] England was not immune to this irresistible desire to breed countless pear varieties, and the Horticultural Society of London listed no less than six hundred pears in their 1826 catalog.[6]

Pear varieties were also imported into America with the help of European settlers, who spread the European pear through seeds taken from cherished trees on their ancestral homelands, which were planted around their village settlements.[7] In 1872 there were one thousand varieties of pears listed in fruit publications throughout the United States, which may seem excessive, but not when compared with the *five* thousand varieties found throughout European literature![8] U. P. Hedrick, the king of American pomology, noted that the number of varieties in the 1920 nursery catalogs amounted to the paltry total of sixty, but that is still more than what is currently available in most contemporary specialty fruit nurseries' mail-order catalogs.[9]

In the United States the cultivation of European pears can suffer from extreme climate fluctuations and from bacterial problems brought on by environmental growing conditions. These factors can make pears an unreliable crop from year to year, but we still believe that pear trees are worth the effort. When the environmental conditions are just right and you are blessed with a good growing season, the reward of trees with a bounty of ripe pears can be one of the greatest treats on Earth, and will help you forget all the countless frustrating years when your trees failed to produce anything but new leaves.

Growth Difficulty Rating: 2. Pear trees are tough and grow in many types of soil, but the trees may have disease and pest problems in certain environments.

Taste Profile and Uses: Pears are in the pome category (along with apple, quince, and other fruiting plants), and have been domesticated and hybridized for thousands of years, which has resulted in a wide diversity of shapes, colors, and sizes. The flavor of pears is

The lovely spring flowers.

Young fruit after pollination.

Young fruit developing

'Anjou' pear developing color.

also an example of this cultivated diversity, with fruits that range from crisp and sweet to juicy and buttery, with many combinations between these two basic types.

European pears are one of the few fruits that will achieve the fullest flavor only when they are picked just before being fully ripe and brought indoors to complete the process of ripening. If pears are allowed to mature on the tree, the fruit can develop a bad texture and flavor because the fruits ripen from the inside out. The best way to determine when to harvest the fruit is to lift a pear upward in the palm of your hand and twist it slightly; this should cause its pedicel (stem) to snap, separating the connection between the fruiting spur and the stem.[10] Pears that cannot easily be separated at the fruiting spur are not ready and should be left on the tree and checked every one to three days. After you harvest the pears, bring them indoors to ripen at room temperatures for one to four days. Pears should be picked in small batches over a period of several weeks.[11]

It is also common for individual trees to produce fruit with different levels of quality. This may be influenced by environmental factors such as the amount of rainfall, fertilization, and pests.

Plant Description: Growth is slow for the trees' first few years, but after getting their root systems established, they can grow fast. Our trees have put on 3 feet (91 cm) of growth in a single season. Pears produce dark glossy green leaves that grow up to 4 inches (10.2 cm) in length, which are ovate with serrated edges, and these are alternately arranged on the branch. Mature trees grow into upright or pyramidal shapes with a variable height that reaches anywhere from 10 to 40 feet (3.1–12.2 m) or larger, depending on the cultivar and type of grafting rootstock used. All cultivated pears are grafted onto rootstocks, which are generally rated as being dwarf, semi-dwarf, or standard in size. The rootstock will determine the size of the tree, soil adaptability, fruiting maturity of the plant, disease and pest resistance, as well as the overall longevity of the tree.[12] Trees that are grafted onto standard rootstocks grow about 25 to 40 feet (7.6–12.2 m) tall. Cultivars grafted onto dwarfing rootstocks can reach 8 to 12 feet (2.4–3.7 m), or slightly larger. Depending on the variety and the size of tree planted, many pear trees take seven years or longer before they begin bearing fruit. As the trees mature and begin to yield crops, they often develop a biannual fruiting habit and will only produce fruit every two years.

Flowers: Pear flowers are always produced on fruit spurs, which are stubby ¾- to 2-inch (1.9–5.1) long growths with pointed tips, arranged at right angles on branches. Pear trees display beautiful flowers with white petals that are ½ to 1½ inches (1.3–3.8 cm) in diameter, which are produced in corymbs (clusters arranged in the form of a flat head) on little branchlets (spurs). Trees can produce abundant amounts of blossoms in early spring and can occasionally get damaged by late frosts.

Pollination Requirements: Although there are a few varieties listed as being self-fertile, such as 'Bartlett', almost all pears require two trees for proper fruit-set. Some specialty nurseries sell multigrafted trees (two or more varieties that are grafted onto a single tree) that are capable of cross-pollination and so produce crops. This may be a good solution for a garden with space for only a single tree.

Although most European pears will cross-pollinate each other, the actual flower blossom period on individual cultivars can be quiet variable, and care should be taken to select cultivars that have flowers which open during the same period. Some mail-order nurseries list the flowering sequence of different varieties to provide a clear idea of which are the most compatible for pollination.

Site and Soil Conditions: Pears are said to be able to tolerate heavy clay soil more than other fruiting plants, but they will ultimately perform better in well-drained soil that is rich in organic matter. Pear trees should be planted in full sun with enough space between the trees for good air circulation to reduce foliar diseases.

Hardiness: European pears are rated at zone 4 to 5 (−30 to −20°F/ −34 to −29°C), depending on the particular variety.

Fertilization and Growth: Pear trees require good amounts of fertilization to produce the best crops and will benefit from a yearly application of aged compost or leaf mulch. But do not use manure or any fertilizers with a heavy nitrogen content, because this promotes rapid growth and makes pears susceptible to fire blight (see the "Pests and Problems" section), which can destroy the trees.[13]

Annual Tree Pruning: Mature pear trees that have grown on a planting site for six or seven years often begin to form into a tangled, crowded framework. A spring pruning to remove dead or crowded branches can alleviate this problem. Thin trees where branches are crowded into one another, to allow a 1-foot (31 cm) spacing between branches, with the goal of allowing sunlight to enter trees and shine directly on developing fruit. This will help to ripen the pears, while allowing air to circulate through the trees and reduce bacterial disease problems.

Cultivars: There are many different cultivars of European pears, and many of the popular cultivars like 'Bartlett' that are being sold in local nurseries are susceptible to diseases such as fire blight. Some nurseries and specialty mail-order fruit catalogs may have a larger selection of cultivars. Starting out with trees that produce delicious fruit and that have a track record of being pest- and disease-resistant will give you the best potential of productivity over time. All of the cultivars below have a reputation for disease resistance.

'Magness': A fruit with a buttery flavor that is similar to Comice. Developed by the USDA for resistance to fire blight and insect pests; rated to zone 5, or −20°F (−29°C).

'Moonglow': A high-quality cultivar that produces great fruit with fire blight resistance that is hardy to zone 4, or −30°F (−34°C).

'Potomac': Developed by the USDA at Ohio State University. A high-quality fruit similar in taste to Anjou. It is fire-blight-resistant and hardy to zone 5, or −20°F (−29°C).

Related Species: Almost 150 years ago the *Pyrus* genus also included many fruiting plants that are no longer considered part of this genus but are closely related. These include mountain ash (*Sorbus*), quince (*Cydonia*), chokeberries (*Aronia*), and hawthorns (*Crataegus*).[14]

Propagation: Pear varieties are traditionally propagated through grafting on a suitable pear rootstock, but quince (*Cydonia*) is also a suitable dwarfing rootstock.

Pests and Problems: Fire blight (*Erwinia amylovora*) is one of the major disease problems that affects pear trees. It is caused by bacterial pathogens that enter the flowers and tips of shoots, and thrives in the younger branches of the trees with soft succulent growth. The bacteria will eventually migrate down the stalk, killing the tissues and causing the leaves and young stems to die, and sometimes the entire tree. Pear trees producing excessive growth are more susceptible to this problem.[15] The presence of fire blight is visible when healthy leaves turn a shiny black color as if burned, and the only known remedy is to cut 2 feet (61 cm) below each of the infected areas of the tree, sterilizing the saw with alcohol between cuts to reduce the transmission of the disease. Anthracnose, canker, scab, and powdery mildew are all minor problems affecting pears.

Insects that can cause minor difficulties include the pear psylla, codling moth, and borers. Because different geographical areas experience specific pest problems, it is best to check with your local university agricultural extensions to determine the pests in your area.

Pecan

Carya illinoinensis

Illinois Nuts

> *The heavy demand for desirable varieties of pecans, has led many unscrupulous persons to enter the pecan nursery business in name only, and these people send out anything that even looks like a young pecan tree, give it either the name of a well-established variety or a new one, and sell it at a high price.*
>
> —F. H. BURNETTE, 1902[1]

Pâcan is a sixteenth-century Algonquin word used by Indigenous tribes to describe all of the edible nuts that needed a stone to crack them open, and the pecan (*Carya illinoinensis*) has been widely recognized as a tree of high utilitarian value.[2] Roughly twelve thousand years ago, Native tribal ancestors redistributed the nuts by carrying them north after the end of the last ice age.[3]

Pecans are giant, majestic trees that grow in the rich, fertile soil of riverbanks throughout the vast plains of the southeastern United States and down through Mexico. Pecans are mostly inhabitants of the

South: They are found in thirteen states, from the Mississippi Valley, up the Missouri River, through Kansas, and across a portion of Ohio into western Indiana, in a total area that covers over a million square miles (2.6 million sq km).[4]

In the late 1800s the nuts became a nutritious staple of southern diets, but the commercial trade in wild pecans began only in the mid-1850s, with Louisiana and Texas shipping nuts to northern states. Due to shorter seasons at the northern limit of the wild pecan's forest range, the trees produce smaller nuts with thicker shells, but farther south the nuts are larger with thinner shells and are thus more valuable as an agricultural crop.[5] The value of pecan trees in forests soon became evident to states with a large amount of forest acreage. In Henderson County, Kentucky, for instance, there were between 300 and 400 solid acres (121.4–161.9 ha) of wild pecan trees that stood over 100 feet (30.5 m) tall, with trunk circumferences up to 16 feet (4.9 m). Some southern forests were "managed" into orchards by removing any small, unprofitable trees to give an advantage to the giant trees that produced superior nuts.[6] Others who wanted to get in on this hot new industry planted pecan nuts, but the trees do not come true to seed, and after two decades of waiting for their trees to yield riches, they ended up with unproductive orchards of trees with variable or bad-quality nuts, which resulted in a slow route to bankruptcy.[7] Southern growers who knew better grafted and budded superior varieties of trees that produced large nuts with thin shells, and the profitable modern pecan industry arrived in America.[8]

Young tree.

Pecans are now grown commercially only in the southern United States.[9] More and more varieties of grafted trees are proving hardy in northern growing conditions, but the full potential of these trees is still not known. If you have patience and a large, open planting area with a lot of sunlight, grafted pecan trees with a proven record of producing early-ripening nuts would be worth a long-term experiment.

Growth Difficulty Rating: 2. After getting established, a pecan tree is easy to grow, and the chief obstacle to planting pecans is that they can mature into trees that are 100 feet (30.5 m) tall. The next major obstacle is that the nuts require a long, hot growing season to mature. In the northern parts of the United States, it is essential to select grafted varieties that have the greatest cold tolerance with the shortest growing season that can be adapted to your location. Areas that have short growing seasons or lots of cool weather will not produce crops of mature nuts.[10]

Taste Profile and Uses: Pecans are best suited to the mild temperament of the patient gardener because trees may take ten years to generate nuts and another five to yield substantial crops; but on the bright side, your great-great-grandchildren can still be cracking open delicious nuts from your trees because they can continue to produce sizable harvests for a hundred years.[11] Mature pecans often begin to exhibit biennial yields, but this is ameliorated by the fact that old established trees can produce 500 pounds (226.8 kg) of nuts each of those years, although there are records of giant Texas pecan trees producing a whopping 2,200 pounds (997.9 kg) in a single season![12]

The nut kernel ripens inside an elongated, 1- to 1½-inch (2.5–3.8 cm) long, thin, reddish brown shell that is pointed at one end and often marked with irregular black striping; this is covered by a green, fleshy shuck (or husk). A ripe pecan nut is ready to harvest in the fall when the shuck turns from green to brown and splits open to reveal the nut shell beneath it, developing a brown color.

Freshly harvested pecans contain a good amount of moisture and need to dry out for long-term storage or they may become moldy. To do this, lay out the nuts on sheets in a warm room that is 75° to 85°F (24–29°C) and move them around occasionally. They should dry out in two to ten days.[13]

Pecans are a nutrient-dense food that can be used in both savory and sweet recipes.

Plant Description: Wild pecan trees can grow over 100 feet (30.5 m) tall with a crown spread of 40 to 70 feet (12.2–21.3 m). Fully grown, they are magnificent, stately ornamental shade trees. These slow-growing deciduous trees produce rich green leaves that are 7 to 10 inches (17.8–25.4 cm) long with serrated edges, and are alternately arranged along branches in groups of nine, eleven, or seventeen leaflets, measuring 12 to 20 inches (30.5–50.8 cm) long.[14]

Flowers: Pecans are monoecious, producing both male and female flowers on the same tree. The staminate (male) flowers grow in groups of drooping clusters of delicate yellow blossoms that are 2 to 4 inches (5.1–10.2 cm) long, called catkins, arranged toward the ends of one- to two-year-old branches. The pistillate (female) flower clusters emerge from the tips of the spring's new growth. The flowers open up at different times; either the male flower opens first followed by the female or the reverse, depending on the cultivar and the growing season. This can determine the size of a crop.[15]

Pollination Requirements: On some trees the female blossoms open first (protogynous) with slightly smaller flowers, but there are some cultivars where the male flowers open first (protandrous), so it is important to have two compatible grafted varieties to ensure pollination.[16] Pecan trees rely on the wind for pollination, and because male and female flowers open up at different times, it is always best to plant two or more trees together.

Site and Soil Conditions: The growth habits of individual pecan trees vary according to the different environments in which they are planted.[17] Pecans are adaptable, and their southern natural habitat is in alluvial sites along rivers, but the trees tend to grow in dry upland forests in the northern part of their natural range. The most important factor in their survival is the climate of the region where they are planted, where the limiting factor is a long growing season with at least 150 to 180 days, with a large amount of summer heat for the best nut development.[18] Pecans must be grown in full sun and have a preference for rich, deep, and moist but well-drained soil that is amended with lots of organic nutrients.

It is important to plant pecans with their mature height in mind, with about 50 to 75 feet (15.2–22.9 m) between trees to allow them to reach their full potential.[19] Grafted trees are slightly shorter than wild ones and often reach about 75 feet in height.

Hardiness: Pecans feature a range of cold tolerance, and in the northern parts of the United States, it is best to choose those cultivars that have been selected for cold hardiness as well as productivity. The hardiest cultivars are rated zone 5, or to −20°F (−29°C), but the limiting factor in that zone will be the length of the growing season.

Pecan leaves.

Male flowers opening.

Mature pecan tree bark.

Handful of hard-shell pecans.

Fertilization and Growth: Pecans are heavy feeders and should be given compost, leaf humus, or well-aged manure in early spring as the trees come out of dormancy, then once again in early fall before the nuts mature. Pecans are slow to begin bearing fruit; you may wait up to ten years to see any nuts,[20] although many nurseries advertise nut production as early as five to seven years, which is probably wishful thinking. From July through early August, the nuts' size increases rapidly, and during this period of development the trees should be irrigated regularly, or the size of the nuts will be stunted.[21]

Cultivars: Although originally the term *paper-shell* was meant to signify a variety of pecan having a very thin shell, in the early 1900s the nursery trade commonly used this term for all the improved varieties.[22] Recommended cultivars vary by geographic region, so it's important to look for cultivars that are best suited for your specific climatic conditions.

'Colby': Grows quickly into a sturdy tree that bears nuts earlier than most other varieties, and produces crops of meaty, thin-shelled nuts that ripen in late September to early October. Trees are hardy to zone 5.

'Peruque': A variety introduced in Missouri in 1953, with a precocious and productive habit that is hardy to zone 6 (−10°F/−12°C), although some sources list it as zone 5 (−20°F/−29°C). The nuts are medium-sized with thin shells and ripen in the middle of September.

'Warren': From a commercial orchard in Missouri and selected for early-ripening nuts and productivity; hardy to zone 5b (−15°F/−26°C).

Related Species: Pecan is closely related to walnut (*Juglans* spp.) and hickory (*Carya* spp.). Pecans can cross-pollinate with hickory trees and produce a hybrid nut called the hican.

Propagation: Pecan tree cultivars can be grafted onto pecan tree rootstock, but then they can take ten years or more to bear nuts, so it may be better to pay more for larger-sized grafted trees that are best suited to your particular geographic region.

Pests and Problems: Stinkbugs can be a serious problem for the backyard pecan grower, leaving behind bitter-tasting nuts with black spots on the kernels. Squirrels, chipmunks, mice, and birds are all avid fans of pecans and will often take nuts that haven't been fully formed.[23]

Red and White Currants

Ribes rubrum

> *While in some regions its fruit is nauseous and unpalatable, in others it has received commendation for the purposes of a jelly. These contrasts show the currant to be a plant variable in nature.*
>
> —Edward Lewis Sturtevant, 1919[1]

Many people could identify a strig of red currants if they were shown a Dutch still-life painting, with the beautiful fruits draped over a bowl, sparkling like strings of small, red gems, but if those individuals were in the United States, most would not have actually tasted the fruit. The lack of knowledge about these tart dessert fruits may be the result of the US federal laws in the early 1900s banning the growing of currants and all members of the *Ribes* genus, which were erroneously blamed for the deaths of white pine trees due to white pine blister rust. But red currants (*R. rubrum*) do not actually carry this virus, and in 1966 most of the laws prohibiting the

The bountiful flower strigs of 'Red Lake'.

Flower strigs of 'Pink Champagne' currants.

The beautiful fruits of 'Red Lake' currants.

The jewel-like fruits of 'Pink Champagne' currants.

cultivation of currants were rescinded. Whatever the reason for Americas' lack of interest in red and white currants, most of the northern parts of Europe are well aware of this fruit, which has been an important staple in their diets for hundreds of years.

There are over 150 species in the *Ribes* genus, but three main ones have been used for breeding the red and white currants in modern gardens: *R. rubrum*, which is native to northern Europe across to Siberia and Manchuria; *R. sativum*, which is indigenous to temperate regions of western Europe; and *R. petraeum*, which grows wild in the high mountain areas of North Africa and Europe. Since all three species are often referred to as red currant, they are often mistaken with one another, and this confusion is only made worse by the fact that white currants are not considered a separate species but actually a sport (natural mutation) of red currants, where the 'red' fruits have lost most of their pigmentation.[2]

The first countries to cultivate the red and white currant were Holland, Denmark, and the coastal plains of the Baltic States.[3] In 1484 the first written note of the berries appears in the *Mainz Herbarius*, where the red currant is celebrated for its medicinal properties. This period probably marks the beginning for the cultivation of this fruit as a garden herb.[4] The first mention of the red currant as a food is in 1536 by the French author Ruellius, who praises both its ornamental and its culinary attributes. In the early 1600s English settlers brought currant seeds to the colonies in the United States, most likely the 'Red Dutch' and 'White Dutch' currant varieties,[5] and thus began the cultivation of red currant (*R. rubrum*) in the eastern parts of the United States, where over time the plants escaped gardens and naturalized.

Red currants are attractive, moderate-sized bushes that are trouble-free and will provide many gardeners with lovely-looking berries.

Growth Difficulty Rating: 1. These plants thrive in many garden settings with few pest problems.

Taste Profile and Uses: Red and white currants are ¼ to ⅔ inch (0.6–1.7 cm) in diameter. In good growing years a mature bush can be completely covered with fruits, which are produced on long grape-like clusters called *strigs* that hang downward, making picking convenient. Red currants start off white and slowly acquire color as the summer progresses until they mature into translucent jewels varying in color from light red to deep scarlet. Most of the white currant varieties mature into a creamy pale yellow or pink-white color. It is best to wait about three weeks after most of the fruits have filled with color to pick them, so that they can develop the best flavor. Cutting off strigs with whole clusters of fruit is easier than going through the time-consuming job of picking each currant.

Although several varieties of currants are palatable for fresh eating when they are fully ripe, most of the fruits are more tart than sweet and better suited to cooking. Currant fruits make wonderful jelly, pies, drinks, and fruit sauces; red currants in particular retain their distinctive tart flavor and attractive scarlet color when used as a jam.

Plant Description: Currants are medium-sized, round-shaped shrubs that often grow as wide as they are tall, ranging in height from 4 to 6 feet (1.2–1.8 m). Unlike their close relation the gooseberry, currant shrubs have smooth branches without thorns. The 2- to 2½-inch (5.1–6.4 cm) leaves on the bushes are alternately arranged with three to five lobes and appear maple-like, with a rich deep green color. Mature bushes form into multi-stemmed shrubs with a crowded, dense growth habit by sending out new branch shoots each year.

Flowers: The greenish yellow flowers are produced in early spring in groups of ten to twenty, on delicate racemes (clusters) in the form of a strig that can measure in length from 1¼ to 2½ inches (3.2–6.4 cm). They are formed near the base of one-year-old stems and on the spurs of older branch stems.[6]

Pollination Requirements: Red and white currants are self-fertile, but better fruiting occurs when two or more bushes are planted together. The bushes are pollinated by hoverflies, bees, and other insects.[7]

Site and Soil Conditions: This shrub prefers cool areas that get consistent moisture and will produce good amounts of fruit in a part-shade environment that has four to five hours of direct sun. For this reason, currant bushes are sometimes interplanted between trees in orchards. They will also grow in full sun, but the foliage often looks burned by the middle of summer and the bushes may drop their leaves and appear dormant. After this they bounce back to life the following spring and continue to yield berries for years with few problems, as long as they get consistent moisture throughout the growing season. Our gardens experience long, hot, humid summers with less rainfall, which are all factors that probably contribute to our shrubs looking tattered and beat-up, but our plants have produced good crops for over twelve years. However, currants are considered to be cool-climate fruiting plants and do not survive for long in the southern parts of the United States, which experience long, hot summers.

The plants will also grow in bad, infertile soil, but they will be healthier, produce more, and live longer when planted in soil that has been amended with a good amount of compost or leaf mulch.[8] Currants have shallow root systems and benefit from a mulch to keep their roots cool.

Hardiness: Red and white currants are rated as being hardy to zone 3, or to −40°F/−40°C.

Fertilization and Growth: All the plants in the *Ribes* genus are heavy feeders that benefit from a liberal top-dressing of well-aged manure or rich compost, mixed with a little potash from fireplace ashes. If this is done at the beginning of spring each year, it will provide the bushes all they need to stay vigorous. The currant's branches tend to be floppy, and one way to deal with this is to plant bushes along a garden fence or wall in order to tie up some of the branches to keep the shrub upright and allow for easier fruit harvesting. A thick stake or garden post can also be used to prop up droopy bushes.

Cultivars: There are about twenty to thirty cultivars of red and white currants available through the nursery trade, and some may be available at local garden centers or through the mail-order nurseries that specialize in fruiting plants.[9]

'Blanka' (*R. rubrum*): A popular cultivar of white currant from Sweden that produces abundant bunches of fruit.

Unripe red currant berries.

'Pink Champagne' (*R. rubrum*): A cultivated variety from 1823 that is a blend of white and red currant, with a flavor akin to the taste of pomegranate. It is one of the best currants for fresh eating.

'Red Lake' (*R. rubrum*): The most widely cultivated variety of red currant. It produces abundant crops of flavorful fruit and was developed at the University of Minnesota in 1933.

Related Species: Red and white currants are closely related to blackcurrants (*R. nigrum*) as well as all the species of gooseberry (*R. uva-crispa*), both described in their own chapters.

Propagation: Currants can be grown from seeds, but it is much easier to take hardwood cuttings in late autumn or midwinter. These cuttings should be about 5 to 7 inches (12.7–17.8 cm) in length and the diameter of a pencil. They can be potted up for a few years until they establish good root systems.

Training and Pruning Considerations: Although currant bushes live a long time, the stems' most productive period begins to wane after four years of bearing fruit. After an established bush reaches six years of age, the best way to remedy this loss in fertility is to selectively cut away some of the older stems toward the base of the plant, which will allow new shoots to spread around the bush, and if this is done over a period of years, a bush can be kept more productive. Different cultivars seem to vary in terms of how vigorously they grow, with some producing robust growth that gets crowded with stems, while others are stingy and concentrate their energy into a few stems. Care should be taken to cut off fewer branches when pruning thinner, skimpier bushes.

Pests and Problems: Currants are generally pest-free but can be bothered by a few problems such as currant aphids (*Cryptomyzus ribis*), currant borers (*Synanthedon tipuliformis*), and powdery mildew (*Podosphaera mors-uvae*). The worst pest that affects bushes is the currant worm (*Nematus ribesii*), an imported agricultural insect. These pests can defoliate bushes completely by eating the leaves, and defoliated plants will often not produce fruit the season after being attacked. Currant worms look like small green caterpillars and can appear on plants in two batches, at the beginning of summer and toward the end of summer. If you catch them early and pick them off leaves, you can mitigate most of the damage to the plants. Birds and chipmunks also like the fruit, so you may need to protect the ripening berries with bird netting.

Red Raspberry

Rubus idaeus, R. idaeus var. *strigosus, R. idaeus* var. *vulgatu, R. strigosus*

> *We are accustomed to boast of the marvelous progress in all lines of American development. What advance can we show in the improvement of the raspberry? Some, to be sure, but most of it has been mere accident. In looking up the history of varieties it is the same story over and over again—"a chance seedling found growing wild, etc."*
>
> —FRED W. CARD, 1914[1]

The typical raspberry encountered in the plastic clamshell containers of the produce aisle at local supermarkets has a long history that is messier than eating the actual fruit. An industrial creation that has been bred for characteristics such as firmness for shipping, size (bigger = better), and color (redder = ripeness), with little attention given to the actual flavor of the fruit, which often seems like an afterthought. If you have ever eaten a ripe raspberry off a bush, you know that the "product" in a supermarket pales by comparison.

One of the parents of modern supermarket raspberries is the North American red raspberry (*Rubus idaeus* var. *strigosus*), which is found along the edges of swamps, bogs, clearings, and open areas that resulted from fires. The North American raspberry is native to the northern United States and southern Canada and is found from the Pacific Coast across the United States to Virginia.[2] The other parent for modern raspberry cultivars is the European red raspberry (*R. idaeus* var. *vulgatus*), which is said to have been harvested from the wild by ancient Greek fruit lovers over two thousand years ago and is native to Europe, North America, and Asia.[3] The species name of *idaeus* comes from the ancient Greek mythological belief that the gods would travel to Mount Ida to go berrying.[4]

The earliest written record of the sweet fruit dates back to the fourth century, but the raspberry was not cultivated as a major fruit crop until the late sixteenth century. By 1925 U. P. Hedrick listed over 400 cultivated varieties of raspberries.[5] English cultivars were crossed with the North American seedlings, which led to improved disease-resistant cultivars, but raspberries are promiscuous and freely hybridize in farm settings as well as in the wild, leading to a large array of intermediate forms, which makes it difficult for taxonomists to identify the subtle differences between species and subspecies.[6] But in the end it doesn't really matter, because even a mediocre raspberry off a bush still beats the flavor of a store-bought product, anytime.

Growth Difficulty Rating: 2. Raspberry canes are easy to grow and among the most productive fruiting plants, often yielding berries a year after planting. But eventually berry brambles will require some management and pruning to deal with the rampant growth of the canes.

Taste Profile and Uses: The raspberry is not an actual berry but a cluster of drupelets or flesh-covered seeds around a receptacle. In contrast with the blackberry, the torus (receptacle or core) is left behind when the fruit is picked. The berry is around 5⁄8 inch (1.6 cm) long to slightly larger, but there are several bred varieties producing extra-large-sized fruits (for instance 'Titan'). The berries are wonderful to eat fresh off the cane; a fully mature one is sweet, rich, and delectable. You can use a bountiful crop to make incredible desserts, jam, wines, and vinegars. The berries can also be frozen and thawed out for later use.

Numerous studies have shown that raspberries contain vitamins A and C, along with various minerals, and are very rich in

Raspberry flower buds.

Flowers and fruit forming.

The highly textured leaf.

Raspberry canes on a simple trellis.

antioxidants.[7] Both the flowers and leaves have been used medicinally for hundreds of years for a wide spectrum of ailments.[8]

Plant Description: A raspberry plant is composed of a series of shoots that sprout from the root system as large arching canes (stems) that grow between 3 and 12 feet (0.9–3.7 m) in length. The stems start out light green then mature into a deep red-brown color and are covered with thin, sharp thorns. The canes produce compound groups of light green leaves with serrated edges in groups of three to five leaflets, which have silvery white undersides. Red raspberries tend to be more cold-hardy and to produce more erect (upright) canes than black or purple raspberries.

Life Cycle of Berry Canes: With the exception of fall-fruiting canes (see section on page 296), all berry bramble stems are biennial. In the first year a new cane (primocane) grows from a bud in the roots into a long slender curving stem with leaves but no flowers. In the second year the original cane (floricane) develops a second flush of foliage with flowers that ultimately produce the fruit, and after flowering the floricanes will die at the end of the growing season. Cane plantings often thrive for eight to ten years. After this length of time the root systems begin to decline for several years until finally dying off. The longevity of berry brambles can be lengthened with a yearly addition of well-aged compost or leaf mulch to the soil surrounding the roots. We have had raspberry patches last over a dozen years by giving them yearly amendments of compost.

Flowers: Raspberry blossoms are composed of five white petals above five long, greenish white sepals (leaf-like structures) that form around the center of a small flower filled with a large collection of stamens and pistils. Each of the flower's pistils that gets pollinated can produce a tiny, separate seed that gets covered by juicy pulp, and all of these grow together to form a single edible fruit or drupe.

Pollination Requirements: The cane systems are self-fertile and often very productive, but because of the relatively small size of the fruit, it is generally best to plant several raspberries together to increase fertility and to produce enough berries for fresh eating and for preserves.

Site and Soil Conditions: The brambles produce excellent-quality fruit in many different types of soil, but avoid planting in wet areas. The plants grow best in well-drained soil, amended with rich organic matter. Raspberries can be grown in part shade but produce larger crops with fewer disease problems when grown in full sunlight. Irrigation is important for good plant growth during dry periods and can improve fruit size and yield. Raspberries have shallow root systems, so they do not like competition with weeds for water and nutrients. Don't plant them where any of the nightshade family (tomatoes, potatoes, peppers, or eggplants) have been grown within the past four years, because these plants can carry a harmful disease called verticillium that can also attack raspberries.[9] It is also a good idea to plant cultivated varieties at least 600 feet (182.9 m) away from any wild raspberry or blackberry plants to reduce the risk of diseases from wild plants spreading to your new plants.[10] We realize this may not be practical in small gardens, but try planting them as far apart as possible.

Managing a Raspberry Patch: There are dozens of ways to deal with berry canes that range from allowing the plants to form a wild bramble patch to formally training them. We listed several different strategies for managing berry brambles in the "Blackberry" chapter.

Hardiness: There is a wide range of cold tolerance for different raspberry cultivars, from zone 5, or −20°F (−29°C), to some cultivars that are rated at zone 3, or −40°F (−40°C).

Fertilization and Growth: Berry canes will grow with neglect, but providing them with an annual layer of garden compost, well-aged manure, or leaf mulch in spring will promote vigorous growth, keep the berry canes' roots cool, help suppress weeds, and increase the longevity of the canes.

Cultivars and Types: There are generally two types of red raspberries: summer-bearing and fall-bearing. These are profiled below.

Summer-Fruiting Raspberries: These are the most common types of raspberry in the nursery trade, which grow as primocanes the first season then overwinter and produce berries during the early-summer months of the next season as floricanes. The canes can bear fruit approximately four to five weeks, with cultivars often listed as producing early-, middle-, and late-season fruits.

'Latham': A very productive and disease-resistant plant that is hardy to zone 3. It yields large berries with great flavor, that ripen early in the season.

'Nova': Hardy canes that yield firm, medium- to large-sized berries with good flavor that ripen midseason. Introduced in 1981 in Nova Scotia, Canada, these productive plants are hardy to zone 3.

'Taylor': An older variety from 1935 that ripens mid- to late season, with large, attractive, excellent-tasting fruit. The canes are vigorous and abundant, and hardy to zone 4 (−30°F/−34°C).

Fall-Fruiting (Everbearing) Raspberries: The new shoots grow into berry canes that produce flowers and fruit on one-year-old primocanes. Once the canes reach a mature length toward the end of summer, they begin fruiting at the tips, progressively bearing fruit down the stems of the cane until the fall temperatures turn cold. After fruiting, the canes are cut flush to the ground, and the cycle begins again in spring.

'Caroline': Hardy to zone 3, a widely adaptable and very productive cultivar that bears medium-sized, flavorful fruit from late summer into late fall.

'Heritage': The most popular of all everbearing raspberries, this cultivar produces very sweet fruit starting in August and is hardy to zone 3.

Related Species: Red raspberry is closely related to black raspberry (*R. occidentalis*; see that chapter), as well as all the other species of wild and cultivated blackberries.

Propagation: Raspberries are vigorous growers and produce stolons (runners), which are horizontal underground stems that grow away from the mother plant and start new berry canes. These are easy to dig up in early spring and can be potted in soil for a few seasons until their root systems get larger.

Pests and Problems: Brambles are fairly easy plants to grow, but starting out with disease-resistant varieties that are sited in open light and planted in consistently moist but well-drained soil will alleviate many of the problems before they affect plants.[11] The principal insect pests of this fruit are the raspberry cane borer (*Oberea bimaculata*), raspberry fruitworm (*Byturus rubi*), and Japanese beetles (*Popillia japonica*).

Gray mold (*Botrytis cinerea*) is a fairly common problem due to environmental factors, such as heavy amounts of rain during bloom time or when the berries are ripening, and can lead to moldy fruit. Other common diseases on raspberries are anthracnose (*Elsinoë veneta*), raspberry mosaic disease (Rubus yellow net virus [RYNV]), and verticillium wilt (*Verticillium* spp.).

Schisandra

Schisandra chinensis

Chinese Magnolia Vine, Five Flavor Berry, Wu Wei Zi

Its Chinese name is Wu Wei Zi, which means "five-taste fruit," derived from its mix of sour, sweet, bitter, warm, and salty taste.

—Susanna Lyle, 2006[1]

In the United States the schisandra vine is a relatively unknown commodity in the nursery trade, but in the temperate northeastern parts of Asia, it is a fruiting vine that grows distinctive tangy-tasting fruit that has been widely used for its medicinal properties for centuries. The vines are widely distributed across Asia in northeastern China, Korea, far-eastern Russia, and northern Japan, and schisandra grows in temperate, mixed-deciduous forests as well as in areas that are primarily coniferous.[2] In Japan there are references to the medicinal uses of schisandra in the pharmacopeial writings of the Ainu culture in the eighteenth and nineteenth centuries.[3] In Chinese folklore the schisandra berry is said to "calm the heart and quiet the spirit."

The berries first appear in Chinese medical writings that are more than four thousand years old, where they are prescribed for treating coughs and general malaise. In the eighteenth century they were used to restore energy to the body.[4] Until recently, all of the schisandra fruit sold commercially in China for medicinal uses was harvested from the wild, and when the vines grew into tree canopies the trees were destructively pulled down for the fruits. Thankfully, in 2010 commercial schisandra farms were started in the north and northeastern provinces of China to grow large-scale commercial crops in a more sustainable manner, for sale to markets across Asia.[5]

Because the fruits emit a lime-like smell when they are crushed, schisandra is called *limonnik* in Russia, and is brewed as a tea by Russian hunters and athletes to help with fatigue and to promote physical stamina.[6] Schisandra was taken from its native haunts in eastern Russia and introduced into European botanical gardens in the late 1850s as "magnolia vine" or "Chinese magnolia vine," because the vines produce tiny flowers that resemble small magnolia blossoms. Schisandra is sold

The fruits hiding under the handsome foliage.

in the nursery trade throughout Europe, but only recently has it been celebrated there as a popular medicinal garden plant.[7]

In the United States the schisandra vine has recently become a little more common through mail-order nurseries as a landscaping plant whose edible berries have medicinal properties. The first time we tried schisandra fruit, we thought it tasted like raw lemon peel and wondered why it was so highly regarded. But when we cooked the berries down and added a sweetener to the strained pulp, the resulting juice tasted like a wonderful, tangy lemonade with a beautiful pink color.

The vines not only yield abundant crops of attractive scarlet-colored berries but have few or no pest problems. These two attributes, combined with schisandra's preference for a part-shade environment where very few edible plants will thrive, make this handsome vine a worthy addition to any garden with a partly shaded area.

Growth Difficulty Rating: 1. Schisandra is a vigorous, easy-to-grow vine, but will eventually require a trellis, post, or fence as a climbing structure as it matures.

Taste Profile and Uses: Beautiful ⅜- to ½-inch (1–1.3 cm) scarlet-red berries hang in attractive grape-like clusters that are up to 3½ inches (8.9 cm) long and highly ornamental. Schisandra berries ripen over a period of two to three weeks beginning in the middle of summer. Because the vine can produce vigorous amounts of foliage, the fruit can often be hidden under the plant's thick covering of dark green leaves.

Schisandra fruit is complex and for most people has a tart-sour berry taste. When a sweetener is mixed with schisandra berries that have been crushed or juiced, the taste comes closest to strawberry lemonade. Sweetened schisandra juice could be used to make jelly, added to cocktails, or used in vitamin-rich desserts like schisandra granita. The berries could also be combined with other fruit sauces, replacing citrus to create new flavor combinations. Some Western taste buds might find the flavor of the fresh or dried fruit too sour to eat raw, but we have come to enjoy schisandra's piquant flavor and eat them out of hand; we also mix the dried berries into fruit and nut balls, and include them in granola. The dried berries can be used to make herbal teas, and young leaves can be lightly cooked and used as a vegetable.

Medicinal Uses: In China, schisandra berry is often dried as a way of preserving the fruits so that they can be stored and used as herbal vitamins. The dried fruits are believed by Chinese herbalists to be a high-energy natural snack and herbal drug for a variety of ailments.[8] Schisandra fruit is considered one of the fifty fundamental herbs for use in traditional Chinese medical practice. [9] There are large-scale commercial farms in China dedicated to producing and harvesting schisandra berries for export to Asian consumers and traditional Chinese medicine practitioners. Recently the berries have become popular in Western health food stores; at the time that this book was written, ground schisandra berry powder was selling for as much as twenty dollars a gram as a health food supplement!

Plant Description: This deciduous, non-invasive woody vine needs a support structure or trellis to climb. The vines do not produce tendrils but tangle around structures to grow; if they wind around tree trunks, they can eventually spread up to 20 to 25 feet (6.1–7.6 m) long.[10] In our gardens, the schisandra has spread 14 feet (4.3 m) after eight years of growing. The attractive rich green, elliptical leaves are 2 to 4 inches (5.1–10.2 cm) long with lightly serrated edges and are alternately arranged along the stem; the foliage has a lemony smell when crushed.

Flowers: In spring the small buds are produced on the previous year's growth in the leaf axils. The ½- to ¾-inch (1.3–1.9 cm) flowers are white to pinkish white, with five to nine waxy petals that are lightly fragrant. The flowers are generated in clusters of up to five, which hang on elongated stalks (peduncles) in a drooping manner, with the blossoms facing the ground. Each cluster can generate multiple berries.

Pollination Requirements: Wild schisandra vines are dioecious and require a separate female and male plant for pollination in order to set fruit. The sexual expression of the plant may be difficult to identify unless it is in flower, so you will need to plant several seedling vines together in order to get fruit crops. 'Eastern Prince' is a reliably self-fertile cultivar that regularly sets large quantities of berries; our single vine began producing fruit after three growing seasons. Schisandra vines are pollinated by beetles, bees, and small moths.[11]

Site and Soil Conditions: Schisandra vines prefer well-drained soil in a part-shade environment that mimics their natural habitat in forest clearings. Some plant sources list it as being adaptable to a full-sun growing site, but we moved our plant away from a location that received eight to nine hours of sun, because the leaves got burned and scorched for two growing years, despite being frequently watered. After we'd relocated the plant in an area with

Mature vine trained on a post.

The delicate flowers of schisandra vine.

Abundant crop of ripening fruits.

Strands of attractive berries.

three to six hours of sun exposure, it has thrived and consistently produced fruit.

This plant would be a good choice for a permaculture food forest because it is both a medicinal and food plant that is an adaptable vine that can be trained up on an arbor or tree, and can thrive in about three to five hours of sunlight. Like all forest plants this vine would benefit from a mulch to keep its roots cool, particularly during the hot parts of the summer.

Hardiness: The vines are cold-hardy to zone 3, or to −40°F (−40°C), because they are adapted to surviving in the cold regions of Russia and northern China.

Fertilization and Growth: This vine does not seem finicky about soil and will grow in a variety of soils conditions, but a yearly application of organic compost or decomposed leaf mulch would make the plants more vigorous and increase their productivity.

Cultivars: 'Eastern Prince' is the most important introduced cultivar of schisandra vine. It is reliably self-fertile and very productive. 'Eastern Prince' was developed by the Vladivostok Station of the Vavilov Institute, which is an important Russian agricultural research institution responsible for introducing improved food crops.

Related Species: There are twenty-three different species in the genus *Schisandra* that grow in temperate and subtropical forests.[12] Magnolia vine (*S. rubriflora*) produces attractive reddish orange flowers and produces berries that are similar to *S. chinensis*. It's native to China and India with a hardiness rating to zone 7 (0°F/−18°C).

Propagation: The easiest way to propagate schisandra is to layer sections of the plant into the soil near the base of the vine in spring, then dig them up in fall after they have had sufficient time to root. These can be potted up and kept in pots for two or three growing seasons to increase the size of their root ball, which will improve the new vine's chances of survival when it is planted in a different location.

Sowing seeds taken from schisandra berries is another method of creating more vines, but they need to be scarified (lightly filed or nicked with a blade) and cold-stratified (subjected to cold) over the winter in pots, though the germination rate on potted seeds is sometimes poor.[13] Softwood cuttings taken in summer have also been a successful method of propagating the vines.[14]

Pests and Problems: Schisandra is not affected by any major pests or diseases. We have successfully grown the vine for over ten years and never encountered any problems.

Seaberry

Hippophae rhamnoides

Sallow Thorn, Sea Buckthorn

> *...Numerous torrents converging on to this grassy pocket have thrust out alluvial fans on which grow dense thickets of* Hippophaë rhamnoides, *a very characteristic formation, for this tree grows along all the more sluggish streams of S.E. Tibet, often in dense thickets.*
>
> —FRANK KINGDON WARD, 1913[1]

Since the year 2000 sea buckthorn has been marketed as "seaberry" to avoid the confusion with buckthorns, which are a distantly related group of plants considered invasive in the Northeast.[2] Seaberry is a fruit that has been gathered by Indigenous peoples as a food and medicine for centuries, and has a wide geographic range that includes China, Mongolia, and Russia, in addition to a scattered distribution across Europe. Published records of seaberry date back to the writings of ancient Greek scholars. In Greece, seaberry shrubs were traditionally used as a fodder for horses to promote weight gain along with a shiny coat—the genus name *Hippophae* literally translates as

Young male seaberry flowers.

Blooming male seaberry flowers.

The narrow silvery leaves.

Ripe seaberries.

"shiny horse."[3] In Tibet there are references to the medicinal uses of seaberry from the eighth century in a Tibetan text called *Gyud Bzi*, or "The Four Books of Pharmacopoeia."[4] In fact many cultures have used the berries for a host of medicinal and culinary uses. Have you ever wondered how the ancient inhabitants of the northern parts of Mongolia and Siberia got healthy doses of vitamins and minerals, in a region where the long arctic winter drops dozens of feet of snow each year and the temperature can drop to −40°F (−40°C)? The answer is seaberry, which has large amounts of vitamins A, E, and K, along with one of the highest concentrations of vitamin C of all fruits (in fact the berries have four times the amount of vitamin C as oranges by weight).[5]

The commercial uses of seaberry fruit began in the 1940s in the Soviet Union, which developed products to take advantage of the fruit's health properties. Russian scientists created preparations of the berries as a topical cream for the faces of the cosmonauts in their space program, to protect them against cosmic radiation.[6] Chinese scientists have not been far behind the Russians; at the 1988 Seoul Olympic Games, the official sports drink for the Chinese athletes was made from seaberries.[7] At the end of the twentieth century, most of the production and processing of the berries switched from the Soviet Union to Mongolia and China. Currently China has between 150 and 200 factories that process the fruit.[8]

But these berries are still widely popular in eastern Europe, and on a trip to Ukraine a few years ago, we saw dozens of babushkas outside a well-known food market in Kiev with large cardboard trays that contained plastic cups filled with freshly picked fruit such as Cornelian cherries, chokeberries, and seaberries that they were selling as a cottage industry. We also think these are wonderful fruits, and if planted in the right space, they could provide large quantities of beautiful and healthy berries for a wide range of uses.

Growth Difficulty Rating: 1. Although easy to grow, they have an aggressive growth habit, and with time the shrubs can spread over 20 feet (6.1 m) from the original planting site.

Taste Profile and Uses: The beautiful orange fruits are considered drupes and are ¼ to ⅜ inch (0.6–1 cm) in diameter. They are produced by female plants on woody stems that are two years old or older. The fruits start out light green and gradually turn a light yellow-orange until finally ripening into a deep orange color in late August. Picking the fruit around the thorns can be a problem, so some people cut branches with the fruit attached and freeze them; upon being thawed out, it's easy to dislodge the fruits by shaking the branches.

The berries are generally too tart and astringent to eat raw, but by lightly cooking them in water with a sweetener then straining out the fruit pulp, you are left with a beverage that tastes like citrus blended with passionfruit. This drink is now available commercially throughout the world as a high-priced health juice.

Seaberries are also used for making alcoholic drinks, and the berries can be dried into a powder for nutritional supplements. The berries are so chock-full of vitamins and minerals that they are now viewed as being one of the most beneficial fruits available in northern climates. Several pharmacological products such as skin cream and soaps are derived from the fruit as well as the seed.[9]

In addition to the berries, fresh or dried leaves can be made into a delicious and healthy tea.

Plant Description: Seaberries are deciduous shrubs with attractive grayish green, willow-like leaves with silvery undersides that range in size from ¾ to 2½ inches (1.9–6.4 cm), alternately arranged on spiny stems. The bushes are vigorous and grow into a variety of shapes that range from a shrub, less than 3 feet (91 cm) tall to giant sprawling bushes that reach up to 16 feet (4.9 m) or larger. The plants are dioecious (separate male and female plants) and have a thick growth habit, with branches bearing multiple sharp spines, so be careful when harvesting the berries.

Flowers: The flowers appear in spring just as the leaves are opening. Each blossom is covered by a russet-brown scale. The male flower scales are larger and created in groups of six to eight growing on short catkins. Female flower buds appear in short racemes (flower clusters) in the axils of the leaves along the entire length of branches. Female flowers are produced in groups of two; when opened, they appear as tiny, curled yellow pistils (female organs of a flower).

Pollination Requirements: Seaberry is dioecious, with plants being either male or female, so at least one of each is required in order to produce fruit. Female flowers are pollinated by the wind in early spring; one male bush can pollinate up to eight female plants. Purchase clearly sexed plants and named cultivars from reputable nurseries to make sure that you get both male and female types.

Site and Soil Conditions: These plants must be in full sunlight or the bushes will decline and ultimately die. Once established, they have good drought tolerance, but we have noticed that seaberry plants must be consistently watered for their first season; after this they are more resilient. Because the shrubs are native to the sandy coastal areas of Europe, they are very tolerant of wind, salt, and cold temperatures and can be effectively planted as a privacy hedge and windbreak. Seaberry has also been used as a decorative hedge in such diverse places as France, the Netherlands, Canada, and Germany because of the root system's ability to spread out and fix nitrogen in sterile soils and to control erosion.[10]

Hardiness: The seaberry is hardy to zone 3, or −40°F (−40°C).

Fertilization and Growth: Initially we had read that seaberry does not like rich soil and thrived best in poor, nutrient-deficient areas.[11] Since the plants' roots fix nitrogen in soil, they should not get any additional nitrogen such as manure around their roots, but they seem to grow more vigorously when mulched with non-nitrogen fertilizers such as compost or decayed leaf mulch; these can be applied every two or three seasons. Our male seaberry bushes tend to "stay put" and send out shoots that are close to the original shrub, but female plants seem almost unstoppable, sending out suckers and forming new bushes in any part of the soil that is in full sun. It is not unusual for female plants to produce root suckers that travel 6 feet (1.8 m) in one season, and over time our female plant has migrated 20 feet (6.1 m) from the original planting site. So, while this is a bad choice for a small garden or yard, seaberry is a great plant for an open area where it can spread into a large edible hedge.

Cultivars: Because of the long historical interest in the health properties of this plant, Russia and eastern Europe have done extensive

Dried female flower parts remaining on young developing fruit.

work in creating higher-quality cultivars selected for superior fruit production, and recently some nurseries have started selling varieties specifically chosen for fruit quality, including those that can be eaten fresh. Though not always found in commerce, there are now over thirty cultivated varieties available through various specialty mail-order nurseries. Here are a few of the cultivars that seem more widely sold.

'Askola': An East German cultivar selected for large fruit and productivity with one of the highest concentrations of vitamin C.

'Baikal': A more compact variety that can be eaten out of hand.

'Orange September': Cultivar with a high sugar content and less thorns than most varieties, making handpicking less painful.

'Titan': A popular and productive tall bush from Belarus that yields good crops of large-sized berries.

Related Species: Chinese sea buckthorn (*Hippophae sinensis*) is a tree that grows 40 to 50 feet (12.2–15.2 m) high with attractive orange fruit that is used like the common seaberry.

Propagation: Seaberry is propagated by seed, softwood cuttings, hardwood cuttings, and layering branches. For female plants the easiest way to create new bushes is by transplanting the large numbers of root suckers that travel under the ground and pop up as "new plants" in places they're not wanted. Dig these up in early spring before the leaves open, and plant them into pots for two or three seasons until they develop large root systems.

Our male plants do not sucker very often and are best propagated by using hardwood cuttings of branches, about 1/4 inch (6 mm) thick and 6 to 8 inches (15.2–20.3 cm) long. The cuttings should be taken in early spring while the plants are still dormant and planted in potting soil in abundant sunlight. These must be kept moist and never allowed to dry out, but not be so wet they rot from standing in water. The best method of achieving this balance is by placing the potted cuttings in a shallow saucer; after the cuttings are watered each day, a small amount of the excess water should always remain in the saucer's bottom for the roots to pull up moisture while the cuttings form roots.

Pests and Problems: Seaberry seems to be fairly pest-free, though it does get attacked every year in our gardens by Japanese beetles (*Popillia japonica*), which only lightly chew on the foliage.

Shipova

Sorbopyrus auricularis

Bollwiller Pear, Bollwyller Pear

> *This intergeneric hybrid is represented by a single clone in arboretums and botanic gardens of Europe and North America, but has not been widely grown for its fruit. The plant has been brought into the United States several times since 1920, from different countries and with different plant names. These introductions appear to be identical.*
>
> —J. D. Postman, 1996[1]

The Rosaceae family contains a large number plants, many of which are familiar, including apple, pear, and quince. On rare occasions some of these distantly related plants can cross-pollinate from one genus to another to create new hybrid species. Whether a random accident of nature or by the hand of a botanist, these cross-pollinations can result in what is called an intergeneric hybrid of two species. The shipova (*Sorbopyrus auricularis*) is an example of a plant that crosses two different genus groups and has received a bewildering number of misleading and hard-to-pronounce Latin names, such as *Bollwilleria auricularis, Azarolis pollvilleriana, Pyrus tomentosa, S. bollwylleriana,* and *S. aucuparia × pyrus*. The source of this confusion is due to the fact that shipova is a combination of two distinct genera, crossing the European pear (*P. communis*) with the whitebeam mountain ash (*S. aria*). This type of hybrid is very rare in nature, and shipova has been almost exclusively propagated by grafts because its fruits produce almost no seeds, or seeds that are usually sterile.[2] However in nature there are no absolutes, and a few shipova seeds actually have produced trees.[3]

Shipova was first mentioned in botanical literature from the early 1400s, then disappeared from history for two hundred years until the Swiss physician and botanist Dominique Chabrey again mentioned the fruits in 1599.[4] He noted "a tall and old tree in the Duke's garden seen from Württemberg in Montbéliard," and went on to say that "it is a beautiful and rare genus of pear that hasn't been seen anywhere else other than Alsace and Montbéliard."[5] Chabrey worked with the botanist Jean Bauhin, who published a major classification of plants called

The small, beautiful fruits.

the *Historia Plantarum Universalis* in 1619 with descriptions of over five thousand plants, including a shipova tree that he saw flower and fruit in 1599, located in the gardens of Bollwyller Castle in Alsace, France.[6] This widely read book became one of the major advances in botanical knowledge in the seventeenth century and introduced more gardeners to the shipova.[7] By the late eighteenth and early nineteenth century these trees had become more commonly known throughout Europe, and gardeners began to cultivate them for their delicious fruits.[8]

Shipovas didn't arrive in the US until 1959,[9] when the first trees were imported from Yugoslavia, but many written sources incorrectly listed their origin as Russia perhaps because there is a Russian town named Shipova.[10] Modern genetic analysis has shown that all the Yugoslavian trees (and original American trees) were grafts taken from a single tree growing on the grounds of the Paris Museum of Natural History in 1834.[11] Which means that the shipova growing in our gardens is most likely from the same venerable old specimen located in Paris, and there is something very cool about the taste of a "fresh" fruit that has a direct lineage over six hundred years old!

Growth Difficulty Rating: 1. Shipova trees are fairly pest-resistant and easy to grow.

Taste Profile and Uses: The fruit is a pome, growing to about 1½ to 2 inches (3.8–5.1 cm), with either a round or a rounded pear shape. It stays green most of the season until it ripens to a ruddy yellow with some blush markings. The (usually) seedless fruit is purported to taste quite delicious, with a delicate, sweet flavor and rose-like aroma. Shipova can take seven years or longer to start to bear fruit; the tree in our gardens flowered for several years before finally producing its first crop. So after ten years of growing we got our first harvest and found that the few shipovas the tree did produce were quite mediocre, with a texture that was mealy and unpalatable. It's one thing to wait ten years for a tree to produce fruit, but it's another thing when the fruit you eat does not live up to written accolades.

Two years after this disappointment, the tree gave us a second crop of several fruits, which were much better tasting and indeed they had a delicate floral taste and fragrance. Two years later we got another small harvest, and the fruit did have a perfumy quality, with meltingly-sweet flesh. Last year our fruits were stolen by ravenous chipmunks, so from here on out we will protect what little fruit the tree produces and hope for bigger harvests!

It should be obvious that although the shipova produces delicious fruit, it is not a tree that will make an impatient gardener happy. Shipova is for the connoisseur who wants a diverse collection of fruit trees with a fascinating historical lineage and is willing to be patient.

Plant Description: Shipova is a slow-growing deciduous tree that matures into a pyramidal shape. Its elliptical green leaves are 3 to 4 inches (7.6–10.2 cm) in length with lightly serrated edges; the underside is grayish green. In the beginning of the season, the upper and lower surfaces of the leaves are lightly hairy, because of its whitebeam parentage, but as the season progresses the upper portion of the leaf loses the downy hairs while the underside remains lightly woolly.

Even though shipova has been cultivated for at least six hundred years, it is still a relatively unknown tree and there is a lot of confusion about the tree's mature size. Some sources list these trees as reaching 20 to 60 feet (6.1–18.3 m), although the contemporary mail-order nurseries that are grafting these trees suggest 12 to 15 feet (3.7–4.6 m). [12] After fifteen years in our gardens, our tree has grown only 10 feet (3.1 m) tall, but this may be explained by our tree being grafted on a dwarf pear rootstock.

Flowers: White downy buds are another nod to the whitebeam parentage and are produced in clusters on short spurs and branches that are about 6 inches (15.2 cm) long.[13] The ¾- to 1-inch (1.9–2.5 cm) white blossoms have five petals. The flowers look most similar to pear blossoms, which is one of shipova's other parents. Flowers are borne out of the leaf axils on fairly long pedicels, and their numerous anthers are a lovely coral pink.

Pollination Requirements: The trees are self-fertile and may actually produce fruit without the need for pollination,[14] but several sources suggest that the size of crops may be increased if two shipovas are planted together or if the tree is planted with a pear tree that has a similar blooming period. Bosc, Comice, and Orcas are all pear varieties listed as being compatible with the shipova for pollination.[15]

Site and Soil Conditions: Shipova should be sited in full sun for best fruiting. The trees are adaptable to different soil conditions but will perform best in well-drained soil that is rich in organic matter such as leaf mulch or compost.

Hardiness: There is a wide range of information on the hardiness of the shipova, but most sources list the trees as hardy to either zone 4 or 5, or about −25°F (−32°C) and some list them as zone 3

The young, downy leaves.

Shipova's attractive flowers.

Young fruits forming.

The small, beautiful fruits.

(−40°F/−40°C).[16] This range of cold tolerance may in part be due to what type of rootstock is being used to graft the trees.

Fertilization and Growth: We deal with our shipova tree much the way we take care of our pear trees. The tree should receive a good amount of fertilization and will produce better-quality fruit from a yearly application of well-aged compost or leaf mulch in the spring. But because these trees have pear lineage, do not use manure or any fertilizers with a heavy nitrogen content—that would promote rapid growth and make the trees more susceptible to fire blight.[17]

Cultivars: There are two cultivars that have been generated from seedlings, although almost all shipova fruit is seedless. One is 'Bulbiformis' (*S. auricularis* var. *bulbiformis*), and the other is *Pyrus* × *malifolia*.[18] Below are a few named cultivars of shipova, but these can sometimes be hard to find even through specialty mail-order fruit catalogs.

'Baby Shipova': A rediscovered older dwarf cultivar of shipova that grows 6 to 8 feet (1.8–2.4 m) in height and supposedly fruits even sooner than the 'Dwarf Shipova'.

'Dwarf Shipova': A shipova grafted onto Aronia rootstock that produces a shorter tree growing 8 to 10 feet (2.4–3.1 m), and has a reputation for bearing fruit within three years.

'Tatarova': A seedling of the original Bollwiller pear produced in Prague in the early twentieth century, with a reputation for producing excellent-quality, small, honey-sweet fruit.[19]

Related Species: Since shipova is a cross between two related genera, the trees have many edible relatives, including the apple (*Malus* spp.), chokeberry (*Aronia* spp.), juneberries (*Amelanchier* spp.), and hawthorn (*Crataegus* spp.).

Propagation: The trees produce fruit that almost never contains seeds, and when shipova fruit does have them, they are almost always sterile. So the best way to produce shipova trees is to graft scions onto a rootstock. The rootstock will determine the ultimate height of the mature tree. A wide range of rootstocks are used for grafting the shipova, including pear, mountain ash, quince, and chokeberry.

Pests and Problems: Because shipova has some pear parentage, the trees can be affected by some of the same pests that affect all pear trees, although this seems to occur less often. In sixteen years of growing the shipova, our tree has not been bothered by any major pests or diseases.

Spikenard

Aralia racemosa

American Spikenard, Indian-Root,
Life of Man, Petty-Morel, Spignet

We seemed to be the only people who love this herbaceous perennial enough to devote more than a few sentences to this humble but wonderful plant that not only provides beauty and structure to the woodland garden but is a worthy fruit as well.

—Allyson Levy and Scott Serrano, 2021[1]

Spikenard is a handsome native perennial wildflower that is found in the rich woods of eastern Quebec to Manitoba, south to Virginia, in upland Georgia and Alabama across to Missouri and eastern Kansas. Spikenards prefer to inhabit the cool shaded areas below mixed deciduous forest tree canopy. Their root systems thrive in the moist, fertile soil built up by decades of decayed leaves and take advantage of the friable leaf duff by sending out giant horizontal, rope-like roots that climb around stones and nearby tree roots to slowly form into colonies. When encountering a mature spikenard plant for the first time in the shade of a forest, the impression is that the plant must be a woody shrub due to the fact that an old specimen can grow 4 to 5 feet (1.2–1.5 m) in a single season. Indeed, these native plants are occasionally sold by nurseries as "architectural" decorative garden plants because of their large leaves and big displays of bottle-brush-like flowers that mature into a dark winey purple hue.

There is an extensive history of spikenard plants being used medicinally by Native tribes for various ailments (hence one of its common names, Indian root), including a drink made from the roots said to be good for a stomachache.[2] In fact this plant is more popular for its medicinal properties than for its culinary uses, which is understandable because these plants belong to the family that includes the important medicinal herb ginseng.[3]

The Native American tribes used spikenard in many foods, including soups, and one particular dish that came from the Menomini people was described as being "very fine"—spikenard cooked with wild onion, wild gooseberry, and sugar.[4] All parts of the plant contain a distinct aromatic flavor, which we appreciate, but it is actually the berries that we love, though these are often given very short shrift in foraging field guides. If they are described at all, it is as either not worth eating, not edible, or poisonous, which is incorrect.[5] We are not sure why some of the older field guides say the berries are toxic, because there are other sources that list them as being edible, and it is their distinct, complex flavor, like a berry mixed with licorice, that we really love.[6] Although the fruits are tiny, a large colony of established plants can provide enough berries to allow you to make a wonderful drink that has a rich root-beer-like flavor by cooking down the berries with sweetener in water.

We feel that these plants are worthy of a shaded spot under some trees, and if you end up not liking the berries, then the birds will take care of them for you.

Growth Difficulty Rating: 1. Spikenard is an easy plant to grow in a part- to full-shade environment that gets consistent moisture through natural rain or watering.

Taste Profile and Uses: Almost all the parts of this perennial plant have been used as cooked food. The small drupe (fleshy thin-skinned fruit that contains up to five tiny seeds) is less than ¼ inch (6 mm) in size and starts out light green, turning to scarlet red, until the early fall when they mature to a dark purple that is almost black when ripe. When the berries are produced in large

The lovely heart-shaped leaves.

Flowers after pollination.

The young berries of spikenard.

Spikenard berries.

quantities, they are highly ornamental, as they are displayed in tight clusters above the leaves. The flavor is a unique combination of berry mixed with tangy licorice. Ripe, dark-colored berries are palatable raw but are much better suited to being cooked with water and a sweetener as a berry drink, tea, or root beer.

Although we have been eating the berries straight off the plant for years, there seems to be no written literature about using the berries raw. So out of an abundance of caution, we suggest that the fruits should be cooked in order to make them safe to eat, and we find that the berries have a better flavor this way. The other parts of the plant that are useful for culinary purposes are the roots, which are traditionally used for making root beer, and the stalks, which are often blanched to make their flavor milder and used like a vegetable in various dishes.

A stately woodland perennial.

Plant Description: Spikenard is a shrubby-looking herbaceous perennial that can ultimately reach 3 to 6 feet (0.9–1.8 m) in height and width. The young stems are green and covered with small white hairs, but as they mature they turn purple and develop two to three large, pinnately compound leaves that can reach up to 2½ feet (76 cm) long. Each leaf is sharply pointed and heart-shaped at the base. The 2 to 6 inches (5.1–15.2 cm) long deep green leaves are divided, and further subdivided into nine to twenty-one ovate leaves that are opposite each other along the stem.

Flowers: The tiny ⅛-inch (3 mm) wide whitish green flowers have five triangular petals that turn backward when fully opened. The inflorescences are arranged as racemes (stalked flowers of nearly equal length from the central flower stalk that forms a flat or curved surface); this flower form is what the species is named for (*racemosa*). Although the individual flowers are very small, they can be quite attractive when they are produced in large quantities and float above the large green leaves. Flowers develop in summer on the end of stems and along the leaf axils.

Pollination Requirements: A single plant is self-fertile and freely produces flowers and berries, but because the berries are so small, several plants should be grouped together to increase the number of berries for harvesting. Also, if you're going to use the plants as a vegetable, you'll need several plants to allow for a regular harvest of a few stems from each plant so that enough of each spikenard will be left intact to allow the plant to stay healthy and vigorous.

Site and Soil Conditions: These are tough, adaptable wildflowers that are native to rocky areas and mountainsides. They can grow in a wide range of soil conditions as long as the site is well drained. The best soil for these plants would be one with plenty of amended leaf mulch or compost, essentially mimicking their natural habitat in the forest. Spikenards are at home in the shady forest and forest edges, but they are most productive when they get about two to four hours of sun in the morning or late afternoon. During long periods of drought, spikenard may need supplemental watering or they will prematurely go dormant until the next spring.

In the past we have mistakenly planted spikenards in full-sun exposure. After a long dry period in the middle of summer, the leaves were fried and the fruits dried up, which caused the plants to go dormant until the following spring. There are sources that list these plants as being adaptable to full sun, but this much light exposure will work only if spikenard gets consistent moisture throughout the hot parts of summer.[7] We ended up transplanting our plants to a shaded spot beneath trees, so they now get a few hours of sun at the end of the day when it is cooler.

Hardiness: Spikenard is rated to zone 3, or −40°F (−40°C), and these plants do well in the colder areas on mountains.

Fertilization and Growth: Like all plants that inhabit the forest, spikenard will grow more vigorously if provided an annual top-dressing of leaf mulch, compost, or well-aged manure. Care should be taken when selecting a planting site because spikenard can grow 6 feet (1.8 m) in a single season, completely covering over small plants growing too close to them. The plant's long horizontal roots can slowly colonize an area, so keep this in mind when choosing a location. When seedling plants have been provided with rich soil and adequate moisture for many years, spikenards may grow into a large colony formed by long root rhizomes that can send up new growth 5 feet (1.5 m) or more from the original plant.

Cultivars: At the time of this publication, there are no known cultivars of spikenard.

Related Species: The *Aralia* genus of plants contains about fifty-five species that include perennials, shrubs, vines, and trees.[8] These plants are also related to the popular medicinal root ginseng (*Panax ginseng* and *P. quinquefolius*).

Propagation: The fresh seeds from spikenard will geminate readily if ripe berries are used. In the fall take ripe berries, separate the seeds from the pulp, and pot them up with soil. These must be cold-stratified over the winter and usually open up the following spring.

Since these plants produce long horizontal roots to form new growth, the simplest way to propagate spikenard is to divide up a section of the root system on an old established plant in late fall when the foliage is dying back and the plant is going dormant. Try to carefully dig around the root system away from the main plant and select a thick piece of the root that has a dormant bud. Pot it up in soil for one growing season to allow the roots to recover from being dug up, and plant it out the following spring. We have had to move these plants several times, and small pieces of the roots sometimes break off and are left behind in the original growing area, often resprouting new plants a year later.

Pests and Problems: Our plants have not had any major problems, but birds will eat the berries.

Wintergreen

Gaultheria procumbens

Checkerberry, Teaberry

> *Once I was taking a young lady student from Sweden for a walk through some woods in central Pennsylvania. I picked some Wintergreen berries for her to taste . . . She did, then exclaimed, "There's that American taste again! It is in the chewing gum, candy and toothpaste, and now it is in the plants and trees that grow over here."*
>
> —EUELL GIBBONS, 1962[1]

That distinctive "American taste" is the rich, aromatic flavor of wintergreen, a sharp root-beer-like flavor contained in the foliage and berries of the plant. These humble but attractive creeping vines are a common species throughout the lower parts of eastern Canada and down the East Coast as far as Georgia, and across to midwestern states such as Wisconsin and Minnesota. Wintergreen is a creeping, mat-forming vine that inhabits cool, shaded forest areas, surrounded by many types of understory plants such as moss, mountain laurel, and huckleberry, which all thrive in the dark, acidic soil, created by the tree canopy of pine and oak trees, combined with other hardwoods.

There are over 130 species of plants in the *Gaultheria* genus, evergreen shrubs and creeping vines that are mostly found in subtropical areas of Asia, Australia, New Zealand, and North America from Canada down to Chile.[2] Five species are native to the woods of the United States. In the eighteenth century, the French Canadian court physician Dr. Hughes Gaultier wrote about his love of wintergreen tea and said that he had "... discovered the Canada Tea ... which he created as an excellent drink." The Swedish explorer Peter Kalm visited his friend Gaultier in 1748, and because of the doctor's enthusiasm for this group of plants, Kalm proposed that this genus should be named in honor of his friend. This idea has "botanically stuck," hence the name *Gaultheria*.[3]

Until modern chemistry created synthetic flavors for consumer commodities, many common household products used to contain wintergreen oil as a popular flavor ingredient, such as toothpaste, medical products, soaps, vermouth, perfume, shaving cream, and root beer. But most people associate wintergreen's aromatic flavor with two popular early American candies: wintergreen Life Savers, which were introduced in 1919 by the Life Savers candy company; and Teaberry chewing gum, which was created by the Clark Brothers Chewing Gum Company in 1921.[4]

We love this old-fashioned flavor, frequently utilize this plant, and think that if you like the taste of wintergreen, you should try growing these creeping vines that thrive in the shade where few other food plants can survive.

Growth Difficulty Rating: 1. Wintergreen is an easy plant to grow that is fairly pest-free and will thrive in almost any shady growing area that has acidic soil.

Taste Profile and Uses: Wintergreen berries are an attractive scarlet-red color and ripen in the fall but will persist on plants throughout the winter. The fruits become sweeter after being exposed to some frosts and cold, but by the end of the winter these berries start to dry out and are less palatable. The round fleshy berries are produced near the ends of the stems and grow about ⅜ inch (1 cm) in diameter, and both the fruit and the leaves contain the same aromatic menthol flavor. The berries are mildly sweet with a distinctive flavor. Tender young leaves can be eaten as a snack, but are best used as tea or for brewing root beer. The leaves contain a volatile oil that will dissipate and weaken if boiled too long. For the best results, place the clean leaves into a saucepan with water and bring to a boil, then remove the pan from the heat and allow the leaves to steep for several hours to increase the flavor. The strained liquid tastes like root beer when combined with a sweetener.

Plant Description: Wintergreen is a spreading, low-growing shrub that slowly forms into a low, wide carpet through rhizomes (underground shoots), which spread at a depth of 1 inch (2.5 cm) below the soil level or less. The creeping branches grow stems that are 2 to 6 inches (5.1–15.2 cm) high and bear attractive, deep green, 1- to 2-inch (2.5–5.1 cm) oval or elliptical leaves, with a smooth leathery texture. The leaves are evergreen and acquire a red blush through the winter, and the cold temperatures reduce the amount of the aromatic oils in the plants, so it's better to harvest the foliage for tea and root beer from summer through the late fall. The plant's leaves and berry displays can be used as a wonderful ground cover around taller acid-loving plants such as pine trees, rhododendrons, and huckleberries. Wintergreen grows slowly and its root systems are easily damaged if they are pulled out of the soil, so take care that the plant stays in place when you harvest the leaves.

Flowers: Small, nodding, bell-shaped white or pinkish white blossoms are formed out of five overlapping petals fused at the tips, which are ¼ to ⅜ inch (0.6–1 cm) long and generally produced singly at the base of the leaves in early summer from June to July.

Wintergreen leaves and developing fruit.

Wintergreen leaves.

The beautiful delicate flowers.

Wintergreen berries.

Berries in late fall frost.

Pollination Requirements: The flowers are self-fertile, but a large group of wintergreen should be sited together to produce a substantial harvest of the berries. An added benefit to planting a lot of this slow-growing plant together is that it will provide a large quantity of leaves so that you will not entirely strip the plants of their foliage when harvesting them for tea.

Site and Soil Conditions: The plants require an acidic soil, but average garden soil can be amended with acidic organic materials such as broken-down pine needles or oak leaves, or the duff found below pine and oak trees. Wintergreen is adaptable to sandy soil as well as areas that have a heavy clay content, but the planting site must be well drained because the plant's root systems can rot when they grow in waterlogged conditions.

Wintergreen does well around mosses and likes similar growing conditions that are shady with a consistent amount of moisture throughout the year, although after the plants have been established for a few growing seasons, they are more drought-tolerant. Wintergreen will grow in full shade, but it does not yield large amounts of flowers and fruits in dense shade. The plants will be more productive when they are sited where they can receive a few hours of sun in morning or in the late afternoon, with shade during the hottest part of the day. Bright shade near the edge of a tall tree line that receives lots of filtered light would also be a better choice than deep shade in the middle of a forest.

Hardiness: Wintergreen is rated hardy to zone 3, or to −40°F (−40°C).

Fertilization and Growth: Wintergreen will benefit from a periodic mulch of pine needles to keep its shallow roots cool and to keep the soil acidic. Wintergreen is in the large family called Ericaceae; all the plants in this family like acidic soil conditions and would benefit from being enriched every two or three years with a small amount of a soil amendment, such as pine duff or a commercially produced organic amendment sold in nurseries for fertilizing rhododendrons and blueberries, which are in the same family as wintergreen.

Cultivars: For many years this plant was ignored by the nursery trade; then a decade ago nurseries began to sell wintergreen as an ornamental ground cover for shady areas. Several years ago plant breeders started to create a handful of cultivars that were selected for larger fruit sizes, berry colors, and yields as well as decorative foliage, and we have listed a few of these newer cultivars.

'Berry Cascade': A trademarked productive variety that produces fruits along the entire stem of the vine, instead of just on the ends of stems.

'Cherry Berries': A cultivar selected for extra-large, showy wintergreen berries.

'Winter Fiesta': A newer cultivar that was selected for its bright white wintergreen berries with a pink blush.

Related Species: The wintergreen genus includes four other plants that are native to the United States: *Gaultheria humifusa* (Alpine wintergreen), *G. ovatifolia* (Oregon spicy wintergreen), *G. hispidula* (creeping snowberry), and *G. shallon* (salal). These are all part of the Ericaceae family, which includes over four thousand species distributed all over the world. Wintergreen is related to a diverse group of plants such as cranberry, rhododendron, manzanita, heather, and mountain laurel.

Propagation: Wintergreen can be grown from seeds that are potted up in soil and cold-stratified (exposed to cold) for two to three months to induce them to germinate. The easiest way to generate new plants is to divide old, established wintergreen colonies by gently digging up a small section of the shrub that has some roots attached to it, and potting it up with acidic, well-drained soil for a few seasons until it has grown a larger root system for transplanting.

Another easy method for propagating plants is to layer a section of the branch under a small mound of soil to encourage roots to form; after a season or two, this part can be cut away from the mother plant and potted up.

Pests and Problems: The plants are not bothered by many pests, but the berries do occasionally get taken by wildlife.

Caution: Wintergreen was used as a natural form of pain relief because all of the parts of the plants contain methyl salicylate, which is a natural aspirin-like chemical. Using small amounts of the leaves and berries for brewing drinks is safe, but people who have an allergic reaction to aspirin should exercise caution when using the leaves.

Wintergreen Oil: Commercially produced wintergreen oil is a common herbal medical product used as a natural remedy for sore muscles because of its beneficial anti-inflammatory properties. But this type of oil contains chemically distilled oils whose methyl salicylate is much stronger than the leaves of the plant. It can be used only as a topical skin cream, and is considered highly toxic and dangerous to ingest in any form!

Acknowledgments

To our very patient friend Daniela Rapp, whose encouragement from the beginning of our project helped to make our writing understandable. For our dear friend Eva Egensteiner, whose eyes helped transform our pictorial documentations into dynamic visuals.

A big thank-you to all the growers and other plant experts who provided information for our writing: Michael Dolan from Burnt Ridge Nursery, Roy Brown at Browns Pecan Farm, and Dr. William Reid at Northern Pecans. We also want to thank the akebia lover, Michael Metivier, who supported us and advocated for our book.

No books that have anything useful to offer are written in a vacuum; they draw upon all the great things written before them. This book is no exception. We could not have written this book without the inspiration and knowledge of many wonderful horticultural writers such as U. P. Hedrick, L. H. Bailey, Michael Dirr, Susanna Lyle, W. J. Bean, Fred W. Card, and Lee Reich, among many others. These writers have helped to shape our thinking and never left our side while putting this book together.

APPENDIX 1

Suggested Further Reading

Listed below are some books filled with useful information on food plants, plant explorers, pest control, medicinal plants, and the history of food production.

Agricultural Explorations in the Fruit and Nut Orchards of China. Frank Nicholas Meyer. 1911. USDA Bureau of Plant Industry Bulletin 204. Reprint edition by Primary Source Editions.

American Chestnut: The Life, Death, and Rebirth of a Perfect Tree. Susan Freinkel. 2007. University of California Press.

American Horticultural Society: Pests & Diseases. Pippa Greenwood. 2000. DK Publishing.

Audels Gardeners and Growers Guide, volume 3: *Fine Fruits for Home and Market*. Edward C. Vick. 1928. Theo Audel & Co.

Blackberry Culture. George M. Darrow. 1918. Farmers Bulletin 643. USDA Publishing.

Blueberries, Cranberries and Other Vacciniums. Jennifer Trehane. 2004. Timber Press.

Bush-Fruits. Fred W. Card. 1914. Macmillan Company.

Chinese Medicinal Plants, Herbal Drugs and Substitutes. Christine Leon and Lin Yu-Lin. 2017. Kew Publishing.

Cornell Guide to Growing Fruit at Home. Marsha Eames-Sheavly et al., 2003. Cornell Cooperative Extension.

Cyclopedia of Hardy Fruits. U. P. Hedrick. 1922. Macmillan Company.

Diseases of Trees and Shrubs, second edition. Wayne A. Sinclair and Howard H. Lyon. 2005. Cornell University Press.

Edible Forest Gardens. Dave Jacke with Eric Toensmeier. 2005. Chelsea Green Publishing.

Edible Wild Plants of Eastern North America. Merritt Lyndon Fernald and Alfred Charles Kinsey. 1958. Harper Publishing.

A Field Guide to Edible Wild Plants of Eastern and Central North America. Lee Allen Peterson. 1977. Houghton Mifflin.

Food: An Authoritative and Visual History and Dictionary of the Foods of the World. Waverly Root. 1980. Smithmark Books.

Food Plants of the World: An Illustrated Guide. Ben-Erik van Wyk. 2005. Timber Press.

Frank N. Meyer, Plant Hunter in Asia. Isabel Shipley Cunningham. 1984. Iowa State University Press.

Fruit & Nuts. Susanna Lyle. 2006. Timber Press.

Fruit, Berry and Nut Inventory, 4th Edition. John Torgrimson. 2009. Seed Savers Exchange.

The Fruits and Fruit Trees of Monticello. Peter J. Hatch. 1998. University Press of Virginia.

Fruits of Eden: David Fairchild and America's Plant Hunters. Amanda Harris. 2015. University of Florida Press.

Growing Fruit. Harry Baker. 2004. The Royal Horticultural Society.

Growing Hybrid Hazelnuts: The New Resilient Crop for a Changing Climate. Phillip Rutter,

Susan Wiegrefe, and Dr. Brandon Rutter-Daywater. 2015. Chelsea Green Publishing.

A History of Horticulture in America to 1860. U. P. Hedrick. 1950. Oxford Press. Reprint by Timber Press, 1988.

Jujube Primer and Source Book. Edited by Roger Meyer and Robert R. Chambers. 1998. California Rare Fruit Growers.

A Modern Herbal, volumes 1 and 2. M. Grieve. 1931. Reprint edition by Dover Press 1971.

Nut Growing. Robert T. Morris. 1921. Macmillan Company.

The Organic Gardener's Handbook of Natural Insect and Disease Control. Edited by Barbara W. Ellis and Fern Marshall Bradley. 1996. Rodale Press.

Pawpaw: In Search of America's Forgotten Fruit. Andrew Moore. 2017. Chelsea Green Publishing.

Perennial Vegetables. Eric Toensmeier. 2007. Chelsea Green Publishing.

The Plums of England. H. V. Taylor. 1949. Crosby Lockwood & Son.

The Plums of New York. U. P. Hedrick. 1923. State of New York Publishing.

Quince Culture. William Witler Meech. 1888. Reprint edition by Kessinger Legacy Reprints, 2019.

The Small Fruits of New York. U. P. Hedrick. 1923. State of New York Publishing.

Tree Crops: A Permanent Agriculture. J. Russell Smith. 1953. Devin-Adair Company.

Uncommon Fruits for Every Garden. Lee Reich. 2004. Timber Press.

Vegetables, Herbs & Spices. Susanna Lyle. 2009. Frances Lincoln Limited Publishing.

Where Our Food Comes From: Retracing Nikolay Vavilov's Quest to End Famine. Gary Paul Nabhan. 2011. Island Press.

The World Was My Garden. David Fairchild. 1938. C. Scribner's Sons.

Below are some useful books about plant propagation.

The Bench Grafter's Handbook: Principles and Practice. Brian E. Humphrey. 2019. CRC Press.

Dwarfed Fruit Trees. Harold Bradford Turkey. 1964. Cornell University Press.

The Grafter's Handbook, fourth edition. R. J. Garner. 1979. Oxford University Press.

Grafting and Budding: A Practical Guide for Fruit and Nut Plants and Ornamentals, second edition. Donald McEwan Alexander and William J. Lewis. 2009. CSIRO Publishing.

The Manual of Plant Grafting. Peter T. MacDonald. 2014. Timber Press.

Plant Propagation: Principles and Practice, ninth edition. Hudson T. Hartmann et al. 2017. Prentice Hall.

Plant Propagation: The Fully Illustrated Plant-by-Plant Manual of Practical Techniques. Alan Toogood. 1999. DK Publishing.

The Reference Manual of Woody Plant Propagation, second edition. Michael A. Dirr and Charles W. Heuser, Jr. 2006. Timber Press.

Seed to Seed: Seed Saving and Growing Techniques for Vegetable Gardeners, second edition. Susanne Ashworth and David Cavagnaro. 2002. Seed Savers Exchange.

APPENDIX 2

Mail-Order Nurseries and Seed Catalogs

Below is a list of mail-order nurseries and internet plant catalogs. The nurseries offer a wide variety of plants for sale. They are a good source for locating grafted fruiting plants, rare edible plants, useful medicinal plants, and heirloom vegetable seeds.

Baker Creek Heirloom Seeds
2278 Baker Creek Road
Mansfield, MO 65704
417-924-8917
www.rareseeds.com

Internet sales and a printed catalog of edible heirloom seeds.

Bay Flora
1569 Solano Avenue #428
Berkeley, CA 94707
510-705-1012
www.bayflora.com

Mail-order nursery of rare fruit plants and trees.

Burnt Ridge Nursery
432 Burnt Ridge Road
Onalaska, WA 98570
360-985-2873
www.burntridgenursery.com

Internet sales and a printed catalog with a wide variety of fruiting plants.

Cricket Hill Garden
670 Walnut Hill Road
Thompson, CT 06787
860-283-1042
www.treepeony.com

Internet sales with a focus on tree peonies and rare grafted fruit trees.

Fedco Seeds
PO Box 520
Clinton, ME 04927
270-426-9900
207-426-0090 (catalog requests)
www.fedcoseeds.com

Internet sales catalog of edible plant seeds and fruiting plants.

Forestfarm at Pacifica
14643 Watergap Road
Williams, OR 97544
541-846-7269
www.forestfarm.com

A wide selection of decorative and edible plants.

High Mowing Organic Seeds
76 Quarry Road
Wolcott, VT 05680
866-735-4454
highmowingseeds.com

Internet sales and a printed catalog of heirloom fruit and vegetable seeds.

Hudson Valley Seed Company
4737 US Route 209
Accord, NY 12404
845-204-8769
hudsonvalleyseed.com

Internet sales of heirloom and open-pollinated garden seeds

Johnny's Selected Seeds
955 Benton Avenue
Winslow, ME 04901
877-564-6697
www.johnnyseeds.com

Internet sales and a printed catalog with a wide range of fruit and vegetable seeds and plants.

One Green World
6469 SE 134th Avenue
Portland, OR 97236
877-353-4028
onegreenworld.com

Internet sales and a printed catalog with a wide variety of edible and fruiting plants.

Raintree Nursery
408 Butts Road
Morton, WA 98356
800-391-8892
raintreenursery.com

Internet sales and a printed catalog with a wide variety of edible and fruiting plants.

Richters Herbs
357 Highway 47
Goodwood, ON L0C 1A
Canada
905-640-6677
www.richters.com

Internet sales and a printed catalog with a wide selection of medicinal and edible seeds.

Seed Savers Exchange
3094 North Winn Road
Decorah, IA 52101
563-382-5990
www.seedsavers.org

Internet sales and a printed catalog with a wide selection of vegetable seeds.

Sheffield's Seeds
269 State Route 34
Locke, NY 13092
315-497-1058
sheffields.com

Mail-order selection of rare decorative and edible seeds.

Stark Bro's
PO Box 1800
Louisiana, MO 63353
800-325-4180
www.starkbros.com

Internet sales and a printed catalog with a variety of fruiting and ornamental plants.

Strictly Medicinal Seeds
PO Box 299
Williams, OR 97544
541-846-0872 (fax)
www.strictlymedicinalseeds.com

Internet sales and a printed catalog with a wide selection of medicinal and edible seeds and plants.

Trees of Antiquity
20 Wellsona Road
Paso Robles, CA 93446
treesofantiquity.com

Internet sales and a printed catalog with a wide selection of edible plants.

Notes

General Considerations

1. Agricultural Research Service, "USDA Plant Hardiness Zone Map," USDA, https://planthardiness.ars.usda.gov.
2. George L. Good and Richard Weir, III, *The Cornell Guide for Planting and Maintaining Trees and Shrubs* (Ithaca, NY: Cornell University Cooperative Extension, 2005), 5–6.

Akebia

1. Edgar Anderson, "The Genus *Akebia*," *Bulletin of Popular Information* (later *Arnoldia*) (May 22, 1934), 17.
2. Christine Leon and Lin Yu-Lin, *Chinese Medicinal Plants, Herbal Drugs and Substitutes* (Richmond, Surrey, U.K.: Kew Publishing, 2017), 458–59; Susanna Lyle, *Fruit & Nuts* (Portland: Timber Press, 2006), 54–55; Li Li et al., "*Akebia*: A Potential New Fruit Crop in China," *HortScience* 45, no. 1 (2010): 4–10, https://journals.ashs.org/hortsci/view/journals/hortsci/45/1/article-p4.xml.
3. Linda Inoki, "*Akebi* (Chocolate Vine)," *Japan Times*, October 15, 2008, https://www.japantimes.co.jp/life/2008/10/15/environment/akebi-chocolate-vine.
4. Eric, "Three Leaf Akebia," *My Corner of Japan* (blog), October 26, 2011, http://japanhomestead.blogspot.com/2011/10/three-leaf-akebia.html; Avi Landau, "Foraging for Wild Fruit in Tsukuba—Akebi—Chocolate Vine," *TsukuBlog*, October 4, 2020, http://blog.alientimes.org/2020/10/foraging-for-wild-eats-in-tsukuba-akebi-again.
5. Anderson, "The Genus *Akebia*," 20.
6. Michael Dirr, *Manual of Woody Landscape Plants*, 5th ed. (Champaign, IL: Stipes Publishing, 1998), 81–82.
7. Li Li, "*Akebia*: A Potential," 4–10.
8. "*Akebia quinata*," Missouri Botanical Garden, accessed October 8, 2021, http://www.missouribotanicalgarden.org/PlantFinder/PlantFinderDetails.aspx?kempercode=a149.
9. "*Akebia quinata*," Missouri Botanical Garden.

Almond

1. Robert T. Morris, *Nut Growing* (New York: Macmillan, 1921), 232.
2. Susanna Lyle, *Fruit & Nuts* (Portland: Timber Press, 2006), 343; Ben-Erik van Wyk, *Food Plants of the World* (Portland: Timber Press, 2005), 309.
3. U. P. Hedrick, *A History of Horticulture in America to 1860* (Portland: Timber Press, reprint 1988, originally published in 1920), 384–85; Henry Smith Williams, *Luther Burbank, His Life and Work* (New York: Hearst's International Library, 1915), 222–23; Lyle, *Fruit & Nuts*, 343.
4. "Using Everything the Orchard Grows," Almond Board of California (2019), http://www.almonds.com/why-almonds/growing-good/zero-waste.
5. Morris, *Nut Growing*, 231; U. P. Hedrick, *The Peaches of New York* (Albany: J. B. Lyon, 1917), 11.
6. Ed Perry and G. Steven Sibbett, "Harvesting and Storing Your Home Orchard's Nut Crop," University of California Division of

Agriculture and Natural Resources (1998), 2, https://anrcatalog.ucanr.edu/pdf/8005.pdf
7. Lyle, *Fruit & Nuts*, 344.
8. John Torgrimson, *Fruit, Berry and Nut Inventory*, 4th ed. (Decorah: Seed Savers Exchange, 2009), 310–12.
9. J. Russell Smith, *Tree Crops: A Permanent Agriculture* (New York: Devin-Adair, 1953), 284.
10. Torgrimson, *Fruit, Berry*, 310–12.

American Chestnut

1. Donald Culross Peattie, *A Natural History of Trees of Eastern and Central North America*, 2nd ed. (New York: Bonanza Books, 1950), 189.
2. Donald E. Davis, "Historical Significance of American Chestnut to Appalachian Culture and Ecology," Dalton State College Social Sciences Division; William L. MacDonald, Franklin C. Cech, and Clay Smith, eds., *Proceedings of the American Chestnut Symposium* (Morgantown: West Virginia University, 1978), 1–3; Sandra L. Anagnostakis and Bradley Hillman, "Evolution of the Chestnut Tree and Its Blight," *Arnoldia* (1992), 3–10.
3. "Breeding for Blight Resistance," American Chestnut Foundation, accessed October 8, 2021, https://www.acf.org/science-strategies/tree-breeding; Stacy L. Clark et al., "Eight-Year Blight (*Cryphonectria parasitica*) Resistance of Backcross-Generation American Chestnuts (*Castanea dentata*) Planted in the Southeastern United States," *Forest Ecology Management* 433 (2019), 153–61, https://www.sciencedirect.com/science/article/abs/pii/S0378112718314002; Sandra L. Anagnostakis, "Chestnut Breeding in the United States for Disease Resistance," Connecticut Agricultural Station (2012), 1–6, https://pubmed.ncbi.nlm.nih.gov/30727322/.
4. Ed Perry and G. Steven Sibbett, "Harvesting and Storing Your Home Orchard's Nut Crop," University of California Division of Agriculture and Natural Resources (1998), 7–9, http://homeorchard.ucdavis.edu/8005.pdf.
5. "Dunstan Chestnut History," Chestnut Hill Nursery, accessed October 8, 2021, https://chestnuthilltreefarm.com/learning-center/dunstan-chestnut-history-2.

American Persimmon

1. Harriet Louise Keeler, *Our Native Trees and How to Identify Them* (New York: Charles Scribner's Sons, 1900), 196.
2. Stephen A. Spongberg, "Notes on Persimmons, Kakis, Date Plums, and Chapotes," *Arnoldia* 39, no. 5 (1979), 290–3.
3. Frank Nicholas Meyer, *Agricultural Explorations in the Fruit and Nut Orchards of China*, USDA Bulletin 204 (1911), 10–16.
4. Spongberg, "Notes on Persimmons," 299; Lee Reich, *Uncommon Fruits for Every Garden* (Portland: Timber Press, 2004), 101–02.
5. Spongberg, "Notes on Persimmons," 293.
6. Spongberg, "Notes on Persimmons," 295.
7. Michael Dirr, *Manual of Woody Landscape Plants*, 5th ed. (Champaign, IL: Stipes Publishing, 1998), 332.

Arctic Kiwi

1. W. J. Bean, *Trees and Shrubs Hardy in the British Isles*, vol. 1 (London: John Murray, 1929), 163.
2. Roger Phillips and Martyn Rix, *The Botanical Garden* (Buffalo: Firefly Books, 2002), 188.
3. Gary L. Koller, "Kolomikta Kiwi" *Arnoldia* (1990), 36–40, http://arnoldia.arboretum.harvard.edu/pdf/articles/1990-50-1-kolomikta-kiwi.pdf .
4. Koller, "Kolomikta Kiwi," 36.
5. James Herbert Veitch, *Hortus Veitchii* (New York: Cambridge Press, 1906), 81, 356.
6. Koller, "Kolomikta Kiwi," 36–40; Lee Reich, *Uncommon Fruits for Every Garden* (Portland: Timber Press, 2004), 68–86.
7. Susanna Lyle, *Fruit & Nuts* (Portland: Timber Press, 2006), 49.

8. Koller, "Kolomikta Kiwi," 38.
9. Lyle, *Fruit & Nuts*, 49.
10. Koller, "Kolomikta Kiwi," 38.
11. John Torgrimson, *Fruit, Berry and Nut Inventory*, 4th ed. (Decorah: Seed Savers Exchange, 2009), 291; Reich, *Uncommon Fruits*, 86; Koller, "Kolomikta Kiwi," 38.
12. Lyle, *Fruit & Nuts*, 50.

Asian Pear

1. Frank Nicholas Meyer, *Agricultural Explorations in the Fruit and Nut Orchards of China*, USDA Bulletin 204 (1911), 25.
2. Meyer, *Agricultural Explorations*, 27.
3. Peter Del Tredici, "The Sand Pear—*Pyrus pyrifolia*," *Arnoldia* (2010), 28–29.
4. Peter Blackburne-Maze, *Fruit: An Illustrated History* (Buffalo: Firefly Books, 2002), 74; Susanna Lyle, *Fruit & Nuts* (Portland: Timber Press, 2006), 369; Lee Reich, *Uncommon Fruits for Every Garden* (Portland: Timber Press, 2004), 188–90; Schuyler S. Korban, *The Pear Genome* (New York: Springer International, 2019), 1–2, 51–52.
5. Tredici, "The Sand Pear," 28–29.
6. John Torgrimson, *Fruit, Berry and Nut Inventory*, 4th ed. (Decorah: Seed Savers Exchange, 2009), 205–08; Reich, *Uncommon Fruits*, 195–96; Meyer, *Agricultural Explorations*, 25–30.
7. Barbara W. Ellis and Fern Marshall Bradley, eds., *The Organic Gardener's Handbook of Natural Insect and Disease Control* (Emmaus, PA: Rodale Press, 1996), 170; Reich, *Uncommon Fruits*, 192.

Beach Plum

1. Richard H. Uva, "Taming the Wild Beach Plum," *Arnoldia* (2003), 11.
2. Michael Dirr, *Manual of Woody Landscape Plants*, 5th ed. (Champaign, IL: Stipes Publishing, 1998), 784.
3. Lee Reich, *Uncommon Fruits for Every Garden* (Portland: Timber Press, 2004), 21; Waverley Root, *Food: An Authoritative and Visual History and Dictionary of the Foods of the World* (New York: Smithmark, 1980), 365.
4. Uva, "Taming the Wild," 13–16.
5. Dirr, *Manual of Woody*, 784; Reich, *Uncommon Fruits*, 22.
6. Edgar Anderson and Oliver Ames, "Botanizing from an Airplane," *Bulletin of Popular Information* (later *Arnoldia*) 3, vol. 7, no. 10–11 (1932), 37–44.
7. Uva, "Taming the Wild," 17.
8. Johnathan Damery, "Recalling Plums from the Wild," *Arnoldia* 75, no. 3 (2018), 24–34.

Black Raspberry

1. U. P. Hedrick, *Cyclopedia of Hardy Fruits* (New York: Macmillan, 1922), 269–70.
2. Hedrick, *Cyclopedia*, 270.
3. Hedrick, *Cyclopedia*, 270; U. P. Hedrick, *The Small Fruits of New York* (Albany: State of New York Publishing, 1923), 153–79.
4. Fred W. Card, *Bush-Fruits* (New York: Macmillan, 1914), 160; Hedrick, *The Small Fruits*, 270; Lori Bushway et al., *Raspberry and Blackberry Production Guide for the Northeast, Midwest, and Eastern Canada* (Ithaca, NY: Natural Resource, Agriculture, and Engineering Service, 2008), 21.
5. Edward C. Vick, *Audels Gardeners and Growers Guide*, vol. 3, *Fine Fruits for Home and Market* (New York: Theo Audel, 1928), 317.
6. Bushway et al., *Raspberry and Blackberry*, 9.
7. Card, *Bush-Fruits*, 106–07; Vick, *Audels Gardeners*, 317.
8. Card, *Bush-Fruits*, 107; Bushway et al., *Raspberry and Blackberry*, 10.
9. Bushway et al., *Raspberry and Blackberry*, 17, 92.

Black Walnut

1. Donald Culross Peattie, *A Natural History of Trees of Eastern and Central North America*, 2nd ed. (New York: Bonanza Books, 1966), 122–23.
2. Robert Morris, *Nut Growing* (New York: Macmillan, 1921), 189–90.

3. Thomas Elias, *The Complete Trees of North America* (New York: Van Nostrand Reinhold, 1980), 269.
4. Merritt Lyndon Fernald and Alfred Charles Kinsey, *Edible Wild Plants of Eastern North America* (New York: Harper & Brothers, 1958), 6, 28, 149.
5. Fernald and Kinsey, *Edible Wild Plants*, 149–50; Joan Parry Dutton, *Plants of Colonial Williamsburg* (Virginia: Colonial Williamsburg Foundation, 1994), 55; Peattie, *A Natural History*, 123.
6. Edward Goodell, "Walnuts for the Northeast," *Arnoldia* (1984), 9.
7. Susanna Lyle, *Fruit & Nuts* (Portland: Timber Press, 2006), 254.
8. Lyle, *Fruit & Nuts*, 252.
9. William Reid et al., "Growing Black Walnut for Nut Production," University of Missouri Center for Agroforestry (2009), https://extension.missouri.edu/media/wysiwyg/Extensiondata/Pub/pdf/agguides/agroforestry/af1011.pdf.
10. Morris, *Nut Growing*, 194; Elias, *The Complete Trees*, 269.
11. Reid, "Growing Black Walnuts," 3; Goodell, "Walnuts for the Northeast," 10–12.
12. Elias, *The Complete Trees*, 266.
13. Goodell, "Walnuts for the Northeast," 10; Reid, "Growing Black Walnuts," 5.

Blackberry

1. Fred W. Card, *Bush-Fruits* (New York: Macmillan, 1914), 224–25.
2. Susanna Lyle, *Fruit & Nuts* (Portland: Timber Press, 2006), 391–92.
3. Card, *Bush-Fruits*, 224–25; U. P. Hedrick, *The Small Fruits of New York* (State of New York Publishing, 1923), 180–242; U. P. Hedrick, *A History of Horticulture in America to 1860* (Portland: Timber Press, 1988), 29, 39.
4. Lyle, *Fruit & Nuts*, 392; M. Grieve, *A Modern Herbal*, vol. 1 (New York: Dover Publications, reprint 1971, originally published in 1931), 108–10.
5. Lyle, *Fruit & Nuts*, 390.
6. George Darrow, *Blackberry Culture*, USDA Farmers Bulletin 643 (1918), 4; Lyle, *Fruit & Nuts*, 392; Edward C. Vick, *Audels Gardeners and Growers Guide*, vol. 3, *Fine Fruits for Home and Market* (New York: Theo Audel, 1928), 307.

Blackcurrant

1. Fred W. Card, *Bush-Fruits* (New York: Macmillan, 1914), 355.
2. U. P. Hedrick, *Cyclopedia of Hardy Fruits* (New York: Macmillan, 1922), 299–300.
3. U. P. Hedrick, *The Small Fruits of New York* (State of New York Publishing, 1923), 253.
4. Anthony Bratsch and Jerry Williams, "Specialty Crop Profile: Ribes (Currants and Gooseberries)," Virginia Cooperative Extension Publication 438-107 (2009).
5. Susanna Lyle, *Fruit & Nuts* (Portland: Timber Press, 2006), 382.
6. John Torgrimson, *Fruit, Berry and Nut Inventory*, 4th ed. (Decorah: Seed Savers Exchange, 2009), 251.
7. Card, *Bush-Fruits*, 340–41.
8. Torgrimson, *Fruit, Berry*, 251–55; Lee Reich, *Uncommon Fruits for Every Garden* (Portland: Timber Press, 2004), 159–61; Hedrick, *Small Fruits*, 302–10.
9. Michael A. Ellis and Leona Horst, "White Pine Blister Rust on Currants and Gooseberries," Ohio State University Extension (2010), https://ohioline.osu.edu/factsheet/HYG-3205.
10. Hedrick, *Small Fruits*, 256–80.
11. Reich, *Uncommon Fruits*, 128.

Boysenberry

1. George Darrow, "Blackberry and Raspberry Improvement," *United States Department of Agricultural Yearbook of Agriculture* (1937), 496, https://naldc.nal.usda.gov/download/IND43893570/PDF.
2. J. G. Vaughan and C. A. Geissler, *The New Oxford Book of Food Plants* (Oxford, U.K.: Oxford University Press, 2009), 88.

3. David Karp, "Boysenberry, A California Treasure," *LA Times*, May 27, 2010.
4. David Elliott, "The Boysenberry Dream," Knott's Berry Farm (2018), https://www.knotts.com/blog/2018/march/3-14-18-the-boysenberry-dream; Oregon Raspberry and Blackberry Commission "Boysenberry" (2008), https://web.archive.org/web/20111021204733/http://www.oregon-berries.com/boysenberry.cfm.
5. Karp, "Boysenberry"; Zack Tawatari, "Once Plentiful Boysenberries All but Extinct in California," Spectrum News (March 26, 2019), https://spectrumnews1.com/ca/la-west/news/2019/03/26/fruit-of-labor.
6. Susanna Lyle, *Fruit & Nuts* (Portland: Timber Press, 2006), 391.
7. John Torgrimson, *Fruit, Berry and Nut Inventory*, 4th ed. (Decorah: Seed Savers Exchange, 2009), 235.
8. Lyle, *Fruit & Nuts*, 391; Lori Bushway et al., *Raspberry and Blackberry Production Guide for the Northeast, Midwest, and Eastern Canada* (Ithaca, NY: Natural Resource, Agriculture, and Engineering Service, 2008), 5.
9. Darrow, "Blackberry and Raspberry," 4; Bushway et al., *Raspberry and Blackberry*, 8.
10. Torgrimson, *Fruit, Berry*, 235.

Che

1. W. J. Bean, *Trees and Shrubs: Hardy in the British Isles*, vol. 1 (London: John Murray, 1922), 441.
2. Susanna Lyle, *Fruit & Nuts* (Portland: Timber Press, 2006), 169–70; Lee Reich, *Uncommon Fruits for Every Garden* (Portland: Timber Press, 2004), 145–50.
3. California Rare Fruit Growers, "Che: *Cudrania tricuspidata*" (1997), https://crfg.org/wiki/fruit/che.
4. California Rare Fruit Growers, "Che."
5. Lee Reich, "Che: Chewy Dollops of Maroon Sweetness," *Arnoldia* 64, no. 1 (2005), 32.
6. California Rare Fruit Growers, "Che."
7. Reich, "Che: Chewy Dollops," 148–49.
8. Michael Judd, personal correspondence, 2017.
9. Reich, *Uncommon Fruits*, 149.

Chinese Kiwi

1. L. H. Bailey, *The Standard Cyclopedia of Horticulture*, vol. 2, 3rd ed. (New York: Macmillan, 1930), 213.
2. Edward Goodell, "Two Promising Fruit Plants for Northern Landscapes," *Arnoldia* (Fall 1982), 121–28; W. J. Bean, *Trees and Shrubs: Hardy in the British Isles*, vol. 1 (London: John Murray, 1929), 162.
3. Goodell, "Two Promising Fruit," 125.
4. Rachel A. Brinkman, "Hidden Gem Among the Vines: *Actinidia arguta*," *Arnoldia* 76, no. 2 (November 2018), 36.
5. Elizabeth Orenstein, Monica Conlin, and Lisa Levine, "Invasive Hardy Kiwi: An Emerging Invasive in the Northeastern United States," Berkshire Environmental Action Team (February 2019).
6. Emily Tepe, "Exploring the Invasiveness Potential of Kiwiberry," University of Minnesota, February 2020.

Chokeberry

1. William Richard Van Dersal, *Native Woody Plants of the United States: Their Erosion-Control and Wildlife Values*, USDA Miscellaneous Publication No. 303 (1938).
2. James W. Hardin, "The Enigmatic Chokeberries (*Aronia*, Rosaceae)," *Bulletin of the Torrey Botanical Club* 100, no. 3 (1973), 178–84.
3. Susanna Lyle, *Fruit & Nuts* (Portland: Timber Press, 2006), 78.
4. Sara Bir, *The Fruit Forager's Companion* (White River Junction, VT: Chelsea Green, 2018), 56; Aronia Eggert, "About Us," accessed October 8, 2021, https://www.aronia.org.pl/about-us/?lang=en.
5. Merritt Lyndon Fernald and Alfred Charles Kinsey, *Edible Wild Plants of Eastern North*

America (New York: Harper & Brothers, 1943), 229.
6. Mark Brand, "*Aronia*: Native Shrubs with Untapped Potential," *Arnoldia* 67, no. 3 (2009), 23.
7. Eldon Everhart, "Aronia—A New Crop for Iowa," Iowa State University Extension, 2009, https://www.extension.iastate.edu/news/2009/mar/110401.htm.
8. Curt Swarm, "Aronia (black chokeberry)—Alternative Crop for Farmers?," *Oskaloosa (IA) Herald*, August 2016, https://www.oskaloosa.com/opinion/columns/aronia-black-choke-berry----alternative-crop-for-farmers/article_1fec57e2-6b20-11e6-9c0a-cb3184b46220.html.
9. Brand, "Aronia: Native Shrubs," 22; Lyle, *Fruit & Nuts*, 78.
10. Beth Berlin and Kathy Zuzek, "Black Chokeberry," University of Minnesota Extension (2018), https://extension.umn.edu/trees-and-shrubs/black-chokeberry.

Cornelian Cherry

1. Charles S. Sargent, "An Early Spring," *Bulletin of Popular Information* 7, no. 1(later *Arnoldia*) (April 11, 1921), 2.
2. Richard Weaver, "The Cornelian Cherries," *Arnoldia* (March 1976), 50–56.
3. Claudia Swan, *The Clutius Botanical Watercolors: Plants and Flowers of the Renaissance* (New York: Harry Abrams, 1998), 79; John Gerard, *The Herball, or Generall Historie of Plantes* (London: John Norton, 1597), 1283, accessed October 8, 2021, https://www.biodiversitylibrary.org/item/178339#page/1312/mode/1up.
4. Michael Dirr, *Manual of Woody Landscape Plants*, 5th ed. (Champaign, IL: Stipes Publishing, 1998), 268–69.
5. Weaver, "The Cornelian Cherries," 50–56.

Cranberry

1. Jennifer Trehane, *Blueberries, Cranberries and Other Vacciniums* (Portland: Timber Press, 2004), 30.
2. Sarah Whitman-Salkin, "Cranberries, a Thanksgiving Staple, Were a Native American Superfood," *National Geographic* (November 28, 2013), https://www.nationalgeographic.com/news/2013/11/131127-cranberries-thanksgiving-native-americans-indians-food-history.
3. L. H. Bailey, *The Standard Cyclopedia of Horticulture*, vol. 2, 3rd ed. (New York: Macmillan, 1930), 87.
4. Bailey, *The Standard Cyclopedia*, 876.
5. K. Afshar et al., "Cranberry Juice for the Prevention of Pediatric Urinary Tract Infection: A Randomized Controlled Trial," *Journal of Urology* 188, suppl. 4 (2012), https://pubmed.ncbi.nlm.nih.gov/22910239; Jarmo Salo et al., "Cranberry Juice for the Prevention of Recurrences of Urinary Tract Infections in Children: A Randomized Placebo-Controlled Trial," *Clinical Infectious Diseases* 54, no. 3 (2012), https://pubmed.ncbi.nlm.nih.gov/22100577; Whitman-Salkin, "Cranberries, a Thanksgiving Staple."
6. Trehane, *Blueberries, Cranberries*, 41.
7. Trehane, *Blueberries, Cranberries*, 35; Cape Cod Cranberry Growers Association, "Where Tradition Meets Innovation," 2019, https://www.cranberries.org/history.
8. George Darrow, *Managing Cranberry Fields*, USDA Bulletin 1401 (1921), 3.
9. Darrow, *Managing Cranberry Fields*, 3.
10. Edward C. Vick, *Audels Gardeners and Growers Guide*, vol. 3 (New York: Theo Audel, 1928), 349.

Elderberry

1. M. Grieve, *A Modern Herbal*, vol. 1 (New York: Dover Publications, reprint 1971, originally published in 1931), original quote from John Evelyn, 269.
2. Grieve, *A Modern Herbal*, vol. 1, 265–78; Ivan Salamon and Daniela Grulova, "Elderberry (*Sambucus nigra*): From Natural Medicine in Ancient Times to Protection Against Witches in the Middle Ages—A Brief

Historical Overview," *HortScience* 1061 (2015), https://doi.org/10.17660/ActaHortic.2015.1061.2.
3. Susanna Lyle, *Fruit & Nuts* (Portland: Timber Press, 2006), 408–10.
4. Patrick Byers, *The Missouri Elderberry Improvement Program* (2016), http://grow.midwest-elderberry.coop/overview/elderstory_byers_6-18-16.pdf.
5. Chad Finn, "Temperate Berry Crops," in *Perspectives on New Crops and New Uses*, edited by Jules Janick (Alexandria, VA: ASHS Press, 1999), 324–34.
6. Joyce Brobst, ed., *Elderberry: An Herb Society of America Essential Guide* (Kirtland, OH: Herb Society of America, 2013), 47.
7. Brobst, *Elderberry*, 39–41.

European Quince

1. U. P. Hedrick, *Cyclopedia of Hardy Fruits* (New York: Macmillan, 1922), 108.
2. Willam Witler Meech, *Quince Culture* (Whitefish, MT: Kessinger Legacy Reprints, 2019), 13–16.
3. Jules Janick and Robert E. Paull, eds., *The Encyclopedia of Fruit & Nuts* (Oxfordshire, U.K.: CAB International, 2008), 634.
4. Peter J. Hatch, *The Fruits and Fruit Trees of Monticello* (Charlottesville: University Press of Virginia, 1998), 127–28.
5. Joseph Postman, "*Cydonia oblonga*: The Unappreciated Quince," *Arnoldia* 67, no. 1 (2009), 2–9.
6. Postman, "*Cydonia oblonga*," 2–9.
7. L. H. Bailey, *The Standard Cyclopedia of Horticulture*, vol. 2, 3rd ed. (New York: Macmillan, 1930), 936.

Flowering Quince

1. Edgar Anderson, *Arnoldia* 4, vol. 3, no. 3 (May 27, 1935), 9.
2. Claude Weber, "Cultivars in the Genus *Chaenomeles*," *Arnoldia* 23, no. 3 (April 5, 1963), 17.
3. Weber, "Cultivars in the Genus," 17.
4. Weber, "Cultivars in the Genus," 24–75.
5. Charles Sargent, *Bulletin of Popular Information* (later *Arnoldia*) (May 5, 1912), 11.
6. U. P. Hedrick, *Cyclopedia of Hardy Fruits* (New York: Macmillan, 1922), 12.
7. Michael Dirr, *Manual of Woody Landscape Plants*, 5th ed. (Champaign, IL: Stipes Publishing, 1998), 214.
8. Weber, "Cultivars in the Genus *Chaenomeles*," 28.
9. Dirr, *Manual of Woody*, 213–15; John Torgrimson, *Fruit, Berry and Nut Inventory*, 4th ed. (Decorah: Seed Savers Exchange, 2009), 230–31.
10. Dirr, *Manual of Woody*, 214.

Goji

1. Paul Gross et al., *Wolfberry: Nature's Bounty of Nutrition and Health* (Middletown, DE: BookSurge Publishing, 2006), 168.
2. Gross et al., *Wolfberry*, 131.
3. Susanna Lyle, *Fruit & Nuts* (Portland: Timber Press, 2206), 265; Gross et al., *Wolfberry*, 155.
4. Ben-Erik van Wyk, *Food Plants of the World: An Illustrated Guide* (Portland: Timber Press, 2005), 234.
5. Gross et al., *Wolfberry*, 19–144; Subhuti Dharmananda, "Lycium Fruit: Food and Medicine," last updated 2007, http://www.itmonline.org/arts/lycium.htm; van Wyk, *Food Plants*, 234.
6. David Karp, "Goji Plays Hard to Get in the US," *LA Times*, August 5, 2009; Christine Leon and Lin Yu-Lin, *Chinese Medicinal Plants, Herbal Drugs and Substitutes* (Richmond, Surrey, U.K.: Kew Publishing, 2017), 633.
7. Gross et al., *Wolfberry*, 19–43, 80–106.
8. van Wyk, *Food Plants*, 234.
9. Lyle, *Fruit & Nuts*, 266; Gross et al., *Wolfberry*, 149.
10. Lyle, *Fruit & Nuts*, 226.
11. Trevor Pemberton, *Edible Shrubs* (Middletown, DE: Plants for a Future, 2019), 13.
12. Karp, "Goji Plays."

Gooseberry

1. Fred W. Card, *Bush-Fruits* (New York: Macmillan, 1914), 357.
2. U. P. Hedrick, *The Small Fruits of New York* (Albany: J. B. Lyon, 1923), 316–18.
3. "Minor Fruits: Gooseberries and Currants," Cornell University Extension (October 2019), http://www.hort.cornell.edu/fruit/mfruit/gooseberries.html; "Currants and Gooseberries," University of Massachusetts Agricultural Extension (2013–14), https://ag.umass.edu/fruit/ne-small-fruit-management-guide/currants-gooseberries; Cheryl Kaiser and Matt Ernst, "Gooseberries and Currants," University of Kentucky College of Agriculture, Food and Environment, Center for Crop Diversitication Crop Profile CCD-CP-6 (2019).
4. "Minor Fruits," Cornell University; Card, *Bush-Fruits.*
5. U. P. Hedrick, *Cyclopedia of Hardy Fruits* (New York: Macmillan, 1922), 307.
6. Lee Reich, *Uncommon Fruits for Every Garden* (Portland: Timber Press, 2004).
7. Hedrick, *The Small Fruits*, 318.

Goumi

1. Fred W. Card, *Bush-Fruits* (New York: Macmillan, 1914), 488.
2. Card, *Bush-Fruits*, 488; U. P. Hedrick, *Cyclopedia of Hardy Fruits* (New York: Macmillan, 1922), 358.
3. John Torgrimson, *Fruit, Berry and Nut Inventory*, 4th ed. (Decorah: Seed Savers Exchange, 2009), 368.
4. Susanna Lyle, *Fruit & Nuts* (Portland: Timber Press, 2006), 188.
5. Jules Janick and Robert E. Paull, eds., *The Encyclopedia of Fruit & Nuts* (Oxfordshire, U.K.: CAB International, 2008), 338.
6. Michael Dirr, *Manual of Woody Landscape Plants*, 5th ed. (Champaign, IL: Stipes Publishing, 1998), 340.
7. Dirr, *Manual of Woody*, 341.
8. "Autumn Olive," Michigan Department of Natural Resources: Invasive Species Best Control Practices (2012), https://mnfi.anr.msu.edu/invasive-species/AutumnOliveBCP.pdf.
9. "*Elaeagnus multiflora*," Missouri Botanical Garden, accessed October 8, 2021, http://www.missouribotanicalgarden.org/PlantFinder/PlantFinderDetails.aspx?taxonid=279925.
10. "*Elaeagnus multiflora*," Missouri Botanical Garden.

Grapes

1. Philip J. Pauly, "Horticulture and the Development of American Identity," *Arnoldia* 63, no. 2 (January 2004), 14.
2. U. P. Hedrick, *The Grapes of New York* (Albany: New York State Agricultural Experiment Station, 1909), 1–2, https://www.gutenberg.org/files/45978/45978-h/45978-h.htm; Susanna Lyle, *Fruit & Nuts* (Portland: Timber Press, 2006), 456.
3. U. P. Hedrick, *A History of Horticulture in America to 1860* (Portland: Timber Press, reprint 1988, originally published in 1920), 209.
4. John Hilty, "Fox Grape," Illinois Wildflowers, last updated April 19, 2020, https://www.illinoiswildflowers.info/trees/plants/fox_grape.html.
5. One Green World, "Grow Your Own Wine," December 21, 2020, https://onegreenworld.com/grow-your-own-wine.
6. Hedrick, *A History of Horticulture*, 157–530; U. P. Hedrick, *Cyclopedia of Hardy Fruits* (New York: Macmillan, 1922), 233–62.
7. "Black Rot of Grapes," Missouri Botanical Garden, accessed October 8, 2021, https://www.missouribotanicalgarden.org/gardens-gardening/your-garden/help-for-the-home-gardener/advice-tips-resources/pests-and-problems/diseases/fruit-spots/black-rot-of-grapes.aspx.

Hardy Orange

1. Michael Dirr, *Manual of Woody Landscape Plants*, 5th ed. (Champaign, IL: Stipes Publishing, 1998), 760–61.

2. Dirr, *Manual of Woody*; Gerald Klingaman, "Plant of the Week: Trifoliate Orange (Hardy Orange)," University of Arkansas (2007), https://www.uaex.edu/yard-garden/resource-library/plant-week/hardy-orange-2-9-07.aspx.
3. Richard F. Lee, "Control of Virus Diseases of Citrus," *Advances in Virus Research* 91 (2015), 143–73.
4. "Trifoliate Orange / Hardy Orange," Trees on the Yale Nature Walk, Yale University (2019), https://naturewalk.yale.edu/trees/rutaceae/poncirus-trifoliata/trifoliate-orangehardy-orange-3.
5. Hong Yu Zhou et al., "Anti-Inflammatory Activity of 21(α, β)-Methylmelianodiols, Novel Compounds from *Poncirus trifoliata* Rafinesque," *European Journal of Pharmacology* 572, no. 2–3 (2007), 239–48, https://doi.org/10.1016/j.ejphar.2007.07.005.
6. John Torgrimson, *Fruit, Berry and Nut Inventory*, 4th ed. (Decorah: Seed Savers Exchange, 2009), 343.
7. Torgrimson, *Fruit, Berry*, 343.
8. Hong Yu Zhou et al., "Anti-Inflammatory," 239–48; Hamilton P. Traub and T. Ralph Robinson, "Improvement of Subtropical Citrus Fruit Crops: Citrus," *United States Department of Agricultural Yearbook of Agriculture* 1589 (1937), 798–99, https://naldc.nal.usda.gov/download/IND43893578/PDF.
9. Siqi Li et al., "The Complete Chloroplast Genome Sequence of *Poncirus polyandra* (Rutaceae), an Endangered Species Endemic to Yunan Province, China," *Mitochondrial DNA Part B* 4, no. 1 (2019), https://www.tandfonline.com/doi/full/10.1080/23802359.2019.1565974.
10. "Trifoliate Orange / Hardy Orange," Texas Invasive Species Institute: Invasive Species List, 2014, http://www.tsusinvasives.org/home/database/poncirus-trifoliata; "2007 Additions to the Alabama Invasive Plant Council's List of Invasive Plants," Alabama Cooperative Extension System, https://www.se-eppc.org/alabama/newadditions.pdf.

Hazelnut

1. Robert T. Morris, *Nut Growing* (New York: Macmillan, 1921), 203.
2. Susanna Lyle, *Fruit & Nuts* (Portland: Timber Press), 2006.
3. Morris, *Nut Growing*, 210.
4. Thomas Molnar, "Corylus" *Wild Crop Relatives*, edited by Chittaranjan Kole (New Brunswick, NJ: Rutgers University, 2011), 15–48, www.researchgate.net/publication/226038600_Corylus.
5. Molnar, "Corylus," 15–48.
6. P. Rutter, S. Wiegrefe, and B. Rutter-Daywarer, *Growing Hybrid Hazelnuts* (White River Junction, VT: Chelsea Green, 2015).
7. Ralph Harmer, "Restoration of Neglected Hazel Coppice," Edinburgh Forestry Commission (2004), https://www.forestresearch.gov.uk/documents/6757/FCIN056.pdf.
8. Jason Fischbach and Lois Braun, "A Production and Economic Model for Hedgerow Hazelnut Production in the Midwestern United States" (2017), https://www.midwesthazelnuts.org/uploads/3/8/3/5/38359971/production_and_economic_model_for_hedgerow_hazelnut_production_v4.pdf; C. F. Lunde, S. A. Mehlenbacher, and D. C. Smith, "Survey of Hazelnut Cultivars for Response to Eastern Filbert Blight Inoculation," *HortScience* 35, no. 4 (July 2000), 729–31; J. Julian, C. Seavert, and J. L. Olsen, "An Economic Evaluation of the Impact of Eastern Filbert Blight Resistant Hazelnut Cultivars in Oregon, USA" *Acta Horticulturae* 845 (2009), 725–32.

Heartnut

1. Andrew Samuel Fuller, *The Nut Culturist: A Treatise on the Propagation, Planting and Cultivation of Nut-Bearing Trees and Shrubs, Adapted to the Climate of the United States* (New York: Orange Judd, 1896), 240.

2. J. A. Neilson, "Some Notes on the Japanese Walnut in North America," *Society of Ontario Nut Growers 21st Annual Report* (1930), 39–45, https://www.songonline.ca/library/articles/japanesewalnut_1930.htm.
3. Neilson, "Some Notes," 39–45.
4. H. L. Crane et al., "Nut Breeding," *United States Department of Agricultural Yearbook of Agriculture* (1937), 859, https://naldc.nal.usda.gov/download/IND43893579/PDF; Fuller, *The Nut Culturist*, 239.
5. Robert T. Morris, *Nut Growing* (New York: Macmillan, 1921), 198; L. H. MacDaniels, "Nut Growing in the Northeastern States," *Arnoldia* 1, no. 9-12 (October 1941), 54.
6. Morris, *Nut Growing*, 198.
7. Morris, *Nut Growing*, 201; Crane, "Nut Breeding," 860.
8. MacDaniels, "Nut Growing," 54.
9. Kathy Dice, "Heartnuts" (January 2017), https://www.redfernfarm.com/index.php/2017/01/11/heartnuts; Edward Goodell, "Walnuts for the Northeast," *Arnoldia* (1984), 15.
10. J. Lee Taylor and Ronald L. Perry, "Growing Nuts," Michigan State University Cooperative Extension Publication 237 (1986), 13.
11. Grimo Nut Nursery, "Heartnut," accessed October 8, 2021, https://www.grimonut.com/index.php?p=Products&category=heartnut; Society of Ontario Nut Growers, "Heartnut, *Juglans ailantifolia* var. *cordiformis*" (2003), https://www.songonline.ca/nuts/heartnut.htm.
12. Society of Ontario Nut Growers, "Heartnut"; MacDaniels, "Nut Growing," 46–48.
13. Goodell, "Walnuts," 15; Neilson, "Some Notes."
14. John Torgrimson, *Fruit, Berry and Nut Inventory*, 4th ed. (Decorah: Seed Savers Exchange, 2009), 319.
15. A. F. Yeager and E. M. Meader, "Breeding Better Fruits and Nuts," University of New Hampshire Experiment Station Bulletin 448 (May 1958), 24.

Highbush Blueberry

1. Frederick V. Coville, "Directions for Blueberry Culture," USDA Bulletin 974 (1921), 23.
2. Jennifer Trehane, *Blueberries, Cranberries and Other Vacciniums* (Portland: Timber Press, 2004), 15.
3. Donald Wyman, "The Highbush Blueberry," *Arnoldia* 2, no. 5 (1942), 29.
4. Coville, "Directions for Blueberry," 2–3.
5. Whitesbog Preservation Trust, "Who We Are," accessed October 8, 2021, http://www.whitesbog.org.
6. Wyman, "The Highbush Blueberry," 30; Coville, "Directions for Blueberry," 4–5.
7. New World Encyclopedia, "Blueberry," accessed October 8, 2021, https://www.newworldencyclopedia.org/entry/Blueberry.
8. Michael Dirr, *Manual of Woody Landscape Plants*, 5th ed. (Champaign, IL: Stipes Publishing, 1989), 1050.
9. Trehane, *Blueberries, Cranberries*, 123.
10. Dirr, *Manual of Woody*, 1048–51.
11. Marsha Eames-Sheavly et al, *The Cornell Guide to Growing Fruit at Home*, Cornell University Cooperative Extension (2003), 81.
12. Trehane, *Blueberries, Cranberries*, 25.
13. Coville, "Directions for Blueberry," 12.
14. George O. Clark, "Blueberry Cages," *Arnoldia* 8, no. 6 (1948), 25–28, http://arnoldia.arboretum.harvard.edu/pdf/articles/1948-8--blueberry-cages.pdf.
15. Eames-Sheavly, *The Cornell Guide*, 82.
16. Trehane, *Blueberries, Cranberries*, 123.

Himalayan Chocolate Berry

1. W. J. Bean, *Trees and Shrubs Hardy in the British Isles*, vol. 2, 5th ed. (London: John Murray, 1929), 22.
2. Michael Dirr, *Manual of Woody Landscape Plants*, 5th ed. (Champaign, IL: Stipes Publishing, 1998), 559; Queensland Government, "*Leycesteria formosa*," *Weeds of Australia* website (2016), https://keyserver

.lucidcentral.org/weeds/data/media/Html/leycesteria_formosa.htm.

3. Jane Powers, "Nature Notes: Pheasant Berry Near the End of Its Blooming Period," *The Times (Ireland)*, October 18, 2018.
4. "*Leycesteria formosa*," Missouri Botanical Garden, accessed October 8, 2021, http://www.missouribotanicalgarden.org/PlantFinder/PlantFinderDetails.aspx?taxonid=278966.
5. Alan Toogood, *American Horticultural Society: Plant Propagation* (New York: DK, 1999), 133.

Honeyberry

1. W. J. Bean, *Trees and Shrubs Hardy in the British Isles*, vol. 2, 5th ed. (London: John Murray, 1929), 40.
2. Susanna Lyle, *Fruit & Nuts* (Portland: Timber Press, 2006), 264; Ernest Small, *North American Cornucopia: Top 100 Indigenous Food Plants* (Boca Raton, FL: CRC Press, 2013), 135–38.
3. Lyle, *Fruit & Nuts*, 264; Small, *North American Cornucopia*, 137.
4. Bob Bors, "Haskap Compatibility, Flowering and Ripening Charts for University of Saskatchewan Varieties," University of Saskatchewan (December 2016) , https://research-groups.usask.ca/fruit/documents/haskap/Haskap bloom ripe charts.pdf; "Haskaps—*Lonicera caerulea*," Northern Hardy Fruit Evaluation Project, North Dakota State University (2011–12), https://www.ag.ndsu.edu/carringtonrec/northern-hardy-fruit-evaluation-project/fruit-index/haskap/haskaps-2013-lonicera-caerulea.
5. Bob Bors, "Haskap Berry Breeding and Production Final Report," University of Saskatchewan (2012), 65–66, https://docplayer.net/48362760-Haskap-breeding-production-final-report-january-2012.html.
6. Small, *North American Cornucopia*, 135.
7. Bors, "Haskap Berry Breeding," 65–66.
8. Bean, *Trees and Shrubs*, 40.
9. Honeyberry USA, "Honeyberry Bloom Chart," accessed October 8, 2021, http://www.honeyberryusa.com/honeyberrybloomtimes.html.
10. Bors, "Haskap Berry Breeding," 32.
11. Honeyberry USA "Honeyberry Bloom."
12. Bob Bors, Ellen Sawchuk, and Jill Thomson, "Mildew & Sunburn in Haskap (Honeyberries)," University of Saskatchewan (December 2016), 1.

Huckleberry

1. Jennifer Trehane, *Blueberries, Cranberries and Other Vacciniums* (Portland: Timber Press, 2004), 182.
2. L. H. Bailey, *The Standard Cyclopedia of Horticulture*, vol. 2, 3rd ed. (New York: Macmillan, 1930), 1320.
3. U. P. Hedrick, *Cyclopedia of Hardy Fruits* (New York: Macmillan, 1922), 324.
4. James Duke, *Handbook of Edible Weeds* (Boca Raton, FL: CRC Press, 1992), 104–5.
5. Minnesota Wildflowers, "*Gaylussacia baccata* (Black Huckleberry)," accessed October 8, 2021, https://www.minnesotawildflowers.info/shrub/black-huckleberry.
6. "Black Huckleberry," Wildflowers of Illinois in Woodlands, accessed October 8, 2021, https://www.illinoiswildflowers.info/savanna/plants/bl_huckleberry.htm.
7. Corey Gucker, "*Gaylussica baccata*," Fire Effects Information System, USDA Forest Service, Rocky Mountain Station (2006), https://www.fs.fed.us/database/feis/plants/shrub/gaybac/all.html.
8. Native Plant Trust, "*Gaylussacia baccata*: Black Huckleberry," Go Botany, accessed October 8, 2021, https://gobotany.nativeplanttrust.org/species/gaylussacia/baccata; New Hampshire Fish and Game, "Karner Blue Butterfly (*Lycaeides Melissa samuelis*)," accessed October 8, 2021, https://www.wildlife.state.nh.us/wildlife/profiles/karner-blue-butterfly.html.
9. Hedrick, *Cyclopedia*, 324.

Jujube

1. Roger Meyer and Robert R. Chambers, ed., *Jujube Primer & Source Book* (Fullerton: California Rare Fruit Growers, 1998), 4.
2. Shengrui Yao, "Past, Present, and Future of Jujubes—Chinese Date in the United States," *Horticultural Science* 48, no. 6 (2013), 672.
3. Isabel Shipley Cunningham, *Frank N. Meyer, Plant Hunter in Asia* (Ames: Iowa State University Press, 1984).
4. Frank Nicholas Meyer, *Agricultural Explorations in the Fruit and Nut Orchards of China*, USDA Bulletin 204 (1911), 35–40; Yao, "Past, Present, and Future," 672–73.
5. Jane Moorman, "NMSU Launches Website Featuring Newly Trademarked Jujube Cultivars," New Mexico State University (May 16, 2018), https://news.nmsu.edu/2018/05/NMSU-launches-website-featuring-newly-trademarked-jujube-cultivars.html; "Jujube," New Mexico State University, accessed October 8, 2021, https://aces.nmsu.edu/jujube/.
6. Christine Leon and Lin Yu-Lin, *Chinese Medicinal Plants, Herbal Drugs and Substitutes* (Richmond, Surrey, U.K.: Kew Publishing, 2017), 720–23.
7. Shengrui Yao et al., "Jujube (*Ziziphus jujuba* Mill.)Flowering and Fruiting in the Southwest United States," *Horticultural Science* 50, no. 6 (2015), 839–46.
8. Meyer and Chambers, *Jujube Primer*, 7; Lee Reich, *Uncommon Fruits for Every Garden* (Portland: Timber Press, 2004), 215.
9. Yao, "Past, Present, and Future," 676.
10. Halfred Wertz, "Cultivation of Jujube in the Eastern U.S.," in Meyer and Chambers, *Jujube Primer*, 10, 11.
11. Meyer and Chambers, *Jujube Primer*, 7.
12. Yao, "Past, Present, and Future," 673.

Juneberry

1. L. H. Bailey, *The Standard Cyclopedia of Horticulture*, vol. 2, 3rd ed. (New York: Macmillan, 1930), 1725.
2. Thomas S. Elias, *The Complete Trees of North America* (New York: Van Nostrand Reinhold, 1980), 592–97.
3. "Juneberries—*Amelanchier alnifolia*," Carrington Research Center, North Dakota State University (2011–17), https://www.ag.ndsu.edu/carringtonrec/archive/northern-hardy-fruit-evaluation-project/fruit-index/juneberry/juneberries-2013-amelanchier-alnifolia; Richard E. Weaver, "The Shadbushes," *Arnoldia* 34, no. 1 (1974), 26.
4. Joyce Newman, "Native Plants 101: The Shadbush Story," New York Botanical Garden (April 25, 2012), https://www.nybg.org/blogs/plant-talk/2012/04/learning/native-plants-101-the-shadbush-story; U. P. Hedrick, *A History of Horticulture in America to 1860* (Portland: Timber Press, reprint 1988, originally published in 1920), 401–02.
5. Michael Dirr, *Manual of Woody Landscape Plants*, 5th ed. (Champaign, IL: Stipes Publishing, 1998), 96.
6. Jim Ochterski, "Consumer Response to Juneberries: Health Benefits Matter," *New York Fruit Quarterly* 18, no. 4 (Winter 2010), 15; Jim Ochterski, "Juneberries—They Go Where Blueberries Can't," Cornell Small Farms Program (October 3, 2011), https://smallfarms.cornell.edu/2011/10/juneberries-they-go-where-blueberries-cant.
7. Cheryl Kaiser and Matt Ernst, "Juneberries," University of Kentucky College of Agriculture, Food and Environment, Center for Crop Diversitication Crop Profile CCD-CP-11 (2019).

Korean Stone Pine

1. Robert T. Morris, *Nut Growing* (New York: Macmillan, 1921), 225.
2. P. Thomas and A. Farjon, "*Pinus koraiensis*," IUCN Red List of Threatened Species (2013); Susanna Lyle, *Fruit & Nuts* (Portland: Timber Press, 2006), 325.
3. "Pine Nuts: Technical Information," International Nut and Dried Fruit Council

(October 2019), https://www.nutfruit.org/files/tech/1572518550_Technical_Information_Kit_Pine_Nuts.pdf; Paul W. Meyer, "The Return to China, Mother of Gardens," *Arnoldia* 68, no. 2 (2010), 9.
4. Christina Harrison and Tony Kirkham, *Remarkable Trees* (London: Thames and Hudson, 2019), 112; "Pine Nuts: Technical Information."
5. Steve Pierson and Robert Lovett, *Pine Cones*, The Lovett Pinetum Charitable Foundation.
6. Lyle, *Fruit & Nuts*, 327.
7. Michael Dirr, *Manual of Woody Landscape Plants*, 5th ed. (Champaign, IL: Stipes Publishing, 1998), 734; John Torgrimson, *Fruit, Berry and Nut Inventory*, 4th ed. (Decorah: Seed Savers Exchange, 2009), 327.
8. Peter MacDonald, *The Manual of Plant Grafting* (Portland: Timber Press, 2014), 123–39; Alan Toogood, *The American Horticultural Society: Plant Propagation* (New York: DK, 1999), 86.
9. "Amur Tiger," IUCN Red List of Endangered Species (February 2021), https://www.iucnredlist.org/species/15956/5333650; "Russia Introduces Ban on Korean Pine Logging," World Wildlife Fund (2010), https://wwf.panda.org/?197032/Russia-introduces-ban-on-Korean-Pine-logging.

Lingonberry

1. L. H. Bailey, *The Standard Cyclopedia of Horticulture*, vol. 2, 3rd ed. (New York: MacMillan, 1930), 3425.
2. Ben-Erik van Wyk, *Food Plants of the World: An Illustrated Guide* (Portland: Timber Press, 2005), 377; Finnish Institute for Health and Welfare, "Picking Lingonberries," *Arctic Lingonberry* website, accessed October 8, 2021, https://www.arcticlingonberry.fi/en/picking.
3. "Lingonberry Jam," *Nordic Diner* website (October 20, 2014), https://nordicdiner.net/lingonberry-jam.
4. Merritt Lyndon Fernald and Alfred Charles Kinsey, *Edible Wild Plants of Eastern North America* (New York: Harper & Brothers, 1958), 316–17.
5. Cathy Heidenreich, "The Lowdown on Lingonberries," Cornell University College of Agriculture and Life Science Cooperative Extension (June 2010), 1–2, https://cpb-us-e1.wpmucdn.com/blogs.cornell.edu/dist/0/7265/files/2016/12/Lingonberries-s7ajxu.pdf; Jennifer Trehane, *Blueberries, Cranberries and Other Vacciniums* (Portland: Timber Press, 2004), 75.
6. Trehane, *Blueberries, Cranberries*, 242–43; Heidenreich, "The Lowdown on Lingonberries," 1.
7. Finnish Institute for Health and Welfare, "Lingonberries Are Nature's Own Vitamin, Mineral and Polyphenol Pills," *Arctic Lingonberry* website, https://www.arcticlingonberry.fi/en/nutritional+value.
8. Trehane, *Blueberries, Cranberries*, 77.
9. Benjamin Cabellero et al., *Encyclopedia of Food and Health* (Waltham, MA: Academic Press, 2016), 364–71; Finnish Institute, "Picking Lingonberries"; Heidenreich, "The Lowdown on Lingonberries," 5.
10. "*Vaccinium vitis-idaea*," Lady Bird Johnson Wildflower Center, https://www.wildflower.org/plants/result.php?id_plant=vavi.
11. D. Tirmenstein, "*Vaccinium vitus-idaea*," Fire Effects Information System, USDA Forest Service, Rocky Mountain Research Station (1991), https://www.fs.fed.us/database/feis/plants/shrub/vacvit/all.html.
12. Lee Reich, *Uncommon Fruits for Every Garden* (Portland: Timber Press, 2004), 60.

Mayapple

1. Merritt Lyndon Fernald and Alfred Charles Kinsey, *Edible Wild Plants of Eastern North America* (New York: Harper & Brothers, 1958), 206.

2. Sasha M. White, "Disjunct Medicine: A History of the (Two) Mayapple(s)," United Plant Savers (February 14, 2017), https://unitedplantsavers.org/disjunct-medicine-a-history-of-the-two-mayapple-s.
3. Euell Gibbons, *Stalking the Wild Asparagus* (Putney, VT: Alan C. Hood, 1962), 126–27; U. P. Hedrick, *A History of Horticulture in America to 1860* (Portland: Timber Press, reprint 1988, originally published in 1920), 20.
4. Marion Blois Lobstein, "History of the Naming of the MayApple," Prince William Wildflower Society (2021), https://vnps.org/princewilliamwildflowersociety/botanizing-with-marion/history-of-the-naming-of-the-mayapple.
5. L. H. Bailey, *The Standard Cyclopedia of Horticulture*, vol. 2, 3rd ed. (New York: MacMillan, 1930), 2726; "Carter's Little Liver Pills" ingredient list, Museum of Health Care at Kingston, Ontario (2021), https://mhc.andornot.com/en/permalink/artifact15192.
6. Arora Rajesh, et al., "Himalayan Mayapple: Traditional Uses, Clinical Indications and Future Prospects," in *Botanical Medicine in Clinical Practice* (Wallingford, U.K.: CABI International, 2008), 71–84.
7. Willem Meijer, "*Podophyllum peltatum*, Mayapple: A Potential New Cash Crop of Eastern North America," *Economic Botany* 28 (1974), 68–72.
8. Hank Becker, "American Mayapple Yields Anti-Cancer Extract," USDA Research Service (July 17, 2000), https://www.ars.usda.gov/news-events/news/research-news/2000/american-mayapple-yields-anti-cancer-extract.
9. Terence M. Laverty, "Plant Interactions for Pollinator Visits: A Test of the Magnet Species Effect," *Oecologia* 89, no. 4 (April 1992), 502–08; "Mayapple," Wildflowers of Illinois in Woodlands, accessed October 8, 2021, https://www.illinoiswildflowers.info/woodland/plants/mayapple.htm.

Maypop

1. Merritt Lyndon Fernald and Alfred Charles Kinsey, *Edible Wild Plants of Eastern North America* (New York: Harper & Brothers, 1958), 275–76.
2. Ronald Boender et al., *Passionflowers: A Pictorial Guide* (independently published, 2019), 1.
3. *Passiflora Online Journal*, https://www.passionflow.co.uk/passiflora-online-journal; L. H. Bailey, *The Standard Cyclopedia of Horticulture*, vol. 3, 3rd ed. (New York: MacMillan, 1930), 2480; W. J. Bean, *Trees and Shrubs Hardy in the British Isles*, vol. 2, 5th ed. (London: John Murray, 1929), 124; Susanna Lyle, *Fruit & Nuts* (Portland: Timber Press, 2009), 312.
4. Steven Foster, "Herbs for Health: Passionflower for Food, Health and Beauty," *Mother Earth Living* (April/May 1997), https://www.motherearthliving.com/gardening/herbs-for-health-passionflower-food-health-beauty; Kristen Johnson Gremillion, "The Development of a Mutualistic Relationship Between Humans and Maypops (*Passiflora incarnata* L.) in the Southeastern United States," *Journal of Ethnobiology* 92, no. 2 (Winter 1989), 135–55.
5. Debbie Orick, "Plant Fact Sheet: Purple Passionflower," USDA Natural Resources Conservation Service (August 2008), https://www.nrcs.usda.gov/Internet/FSE_PLANTMATERIALS/publications/arpmcfs8018.pdf; Shawna Cain, "Purple Passion Flower Used as Food and Medicine," *Cherokee Phoenix*, October 22, 2009.
6. Myles S. Irvine, ed., "Which Passion Fruit Are Edible?", *Passiflora Online Journal* (2009), https://www.passionflow.co.uk/edible-passion-fruit.
7. Irvine, *Passiflora Online*; Lee Reich, *Uncommon Fruits for Every Garden* (Portland: Timber Press, 2004), 140.
8. Reich, *Uncommon Fruits*, 140.

Medlar

1. Edgar Anderson, *The Medlar (Mespilus germanica), Dr. Samuel Johnson's Standard for Wit*, (Chesterfield, MO: Missouri Botanical Garden Bulletin, 1964–65), 53: 11-12, https://www.biodiversitylibrary.org/item/19143#page/1/mode/1up.
2. Quentin Buvelot, *Destillevens van Adriaen Coorte* (The Hague: Waanders Printers, 2008), 25, 37, 87, 89, 91.
3. Buvelot, *Adriaen Coorte*, 25, 37, 87, 89, 91; Julie Berger Hochstrasser, *Still Life and Trade in the Dutch Golden Age* (New Haven, CT: Yale University Press, 2007), 1625; Lee Hendrix and Thea Vignau-Wilberg, *Mira Calligraphiae Monumenta* (Malibu, CA: J. Paul Getty Museum Publications, 1992), 79.
4. Lee Reich, *Uncommon Fruits for Every Garden* (Portland: Timber Press, 2004), 235–32.
5. John Torgrimson, *Fruit, Berry and Nut Inventory*, 4th ed. (Decorah: Seed Savers Exchange, 2009), 374–75.
6. James B. Phipps, *Hawthorns and Medlars (Royal Horticultural Society Plant Collector Guide)* (Portland: Timber Press, 2003), 104–08.
7. Phipps, *Hawthorns and Medlars*, 104–08; James B. Phipps, "*Mespilus canescens*, a New Rosaceous Endemic from Arkansas," *Systematic Botany* 15, no. 1 (1990), 26–32.
8. Reich, *Uncommon Fruits*, 235–32.

Mulberry

1. L. H. Bailey, *Sketch of the Evolution of Our Native Fruits* (New York: Macmillan, 1898), 144.
2. M. Grieve, *A Modern Herbal*, vol. 2 (New York: Dover Publications, reprint 1971, originally published in 1931), 558–59.
3. L. H. Bailey, *Mulberries*, Cornell University Agricultural Experiment Station Bulletin 46 (1892), 227; E. H. Wilson, *A Naturalist in Western China* (New York: Doubleday, Page, 1913), 76–78.
4. U. P. Hedrick, *A History of Horticulture in America to 1860* (Portland: Timber Press, reprint 1988, originally published in 1920), 136; Bailey, *Mulberries*, 128–34.
5. California Rare Fruit Growers, "Mulberry," 1996, https://crfg.org/wiki/fruit/mulberry; J. Swearingen and C. Bargeron, *2016 Invasive Plant Atlas of the United States*. University of Georgia Center for Invasive Species and Ecosystem Health, http://www.invasiveplantatlas.org/subject.html?sub=6050.
6. Christine Leon and Lin Yu-Lin, *Chinese Medicinal Plants, Herbal Drugs and Substitutes* (Richmond, Surrey, U.K.: Kew Publishing, 2017), 554.
7. California Rare Fruit Growers, "Mulberry."
8. Harriet L. Keeler, *Our Native Trees, and How to Identify Them* (Wentworth Press, reprint 2019, originally published in 1900), 253.
9. California Rare Fruit Growers, "Mulberry."
10. Art Fichtner and Camille Goodwin, "Popcorn Disease on Mulberry," AgriLIFE Extension Service, Texas A&M University, https://aggie-horticulture.tamu.edu/galveston/Gardening_Handbook/PDF-files/GH-067--popcorn-disease-on-mulberry.pdf.

Nanking Cherry

1. U. P. Hedrick, *Systemic Pomology* (New York: Macmillan, 1925), 160.
2. R. A. Howard & A. I. Baranov, "The Chinese Bush Cherry: *Prunus tomentosa*," *Arnoldia* 24, no. 9 (September 1964), 81; Frank Nicholas Meyer, *Agricultural Explorations in the Fruit and Nut Orchards of China*, USDA Bulletin 204 (1911), 23–24.
3. U. P. Hedrick, *The Cherries of New York* (Albany: J. B. Lyon, 1915).
4. E. H. M. Cox, *Plant Hunting in China* (London: Collins, 1945), 193; Meyer, *Agricultural Explorations*, 24.
5. Thaddeus McCamant and Sadie Schroeder, "Perennial Fruit: Nanking Cherries," Minnesota Institute for Sustainable Agri-

culture (2018), 66, https://www.misa.umn.edu/publications/perennialfruit.

6. Howard and Baranov, "The Chinese Bush Cherry," 82; Elizabeth Hamilton et al., "Nanking Cherry in the Garden," Utah State University Extension Paper 1580 (2016).
7. McCamant and Schroeder, "Perennial Fruit," 67.
8. Howard and Baranov, "The Chinese Bush Cherry," 84.
9. McCamant and Schroeder, "Perennial Fruit," 68.
10. Howard and Baranov, "The Chinese Bush Cherry," 84–85.
11. Howard and Baranov, "The Chinese Bush Cherry," 85.

Pawpaw

1. Donald Culross Peattie, *A Natural History of Trees of Eastern and Central North America*, 2nd ed. (New York: Bonanza Books, 1950), 288.
2. Andrew Moore, *Pawpaw: In Search of America's Forgotten Fruit* (White River Junction, VT: Chelsea Green, 2015), 5; Desmond R. Lane, "The Pawpaw (*Asimina triloba* (L.) Dunal): A New Fruit Crop for Kentucky and the United States," *Horticultural Science* 31, no. 5 (September 1996), 788, https://www.researchgate.net/publication/241660339_The_Pawpaw_Asimina_triloba_L_Dunal_A_New_Fruit_Crop_for_Kentucky_and_the_United_States.
3. R. Neal Peterson, "Pawpaw (*Asimina*)," *ISHS Acta Horticulture* 290 (1991), 569–602, http://doi.org/10.17660/ActaHortic.1991.290.13.
4. Meadowcroft Rockshelter and Historic Village, accessed October 8, 2021, https://www.heinzhistorycenter.org/meadowcroft/about/; Moore, *Pawpaw*, 7.
5. Joyce Newman, "Native Plants 101: The Shadbush Story," New York Botanical Garden (April 25, 2012), https://www.nybg.org/blogs/plant-talk/2012/04/learning/native-plants-101-the-shadbush-story; U. P. Hedrick, *A History of Horticulture in America to 1860* (Portland: Timber Press, reprint 1988, originally published in 1920), 401–02.
6. David Fairchild, *The World Was My Garden: Travels of a Plant Explorer* (New York: Charles Scribner's Sons, 1938), 434.
7. Lee Reich, *Uncommon Fruits for Every Garden* (Portland: Timber Press, 2004), 44.
8. Michael Judd, *For the Love of Pawpaws* (self-published, 2019), 25–39.
9. Reich, *Uncommon Fruits*, 52.

Pear

1. U. P. Hedrick, *The Pears of New York* (Albany: J. B. Lyon, 1921), 1.
2. Hedrick, *The Pears*, 2.
3. Susanna Lyle, *Fruit & Nuts* (Portland: Timber Press, 2006), 367; Hedrick, *The Pears*, 3.
4. Hedrick, *The Pears*, 14–15.
5. Hedrick, *The Pears*, 16–17.
6. Hedrick, *The Pears*, 37.
7. Hedrick, *The Pears*, 40.
8. U. P Hedrick, *Cyclopedia of Hardy Fruits* (New York: Macmillan, 1922), 76.
9. Hedrick, *Cyclopedia*, 76.
10. E. L. D. Seymour, *The Garden Encyclopedia* (New York: Wm. H. Wise, 1936), 880.
11. Harry Baker, *Growing Fruit (The RHS Encyclopedia of Practical Gardening)* (London: Octopus Publishing Group, 1999), 125.
12. "*Pyrus communis*," Missouri Botanical Garden, accessed October 8, 2021, http://www.missouribotanicalgarden.org/PlantFinder/PlantFinderDetails.aspx?kempercode=a897; "Rootstocks for Pear," Washington State University Comprehensive Fruit Tree Site, accessed October 8, 2021, http://treefruit.wsu.edu/web-article/pear-rootstocks.
13. Barbara W. Ellis and Fern Marshall Bradley, eds., *The Organic Gardener's Handbook of Natural Insect and Disease Control* (Emmaus, PA: Rodale Press, 1996), 70; Fern Marshall Bradley and Barbara W. Ellis, *Rodale's*

All-New Encyclopedia of Organic Gardening (Emmaus, PA: Rodale Press, 1992), 257.

14. L. H. Bailey, *The Standard Cyclopedia of Horticulture*, vol. 2, 3rd ed. (New York: Macmillan, 1930), 2865.
15. Bailey, *The Standard Cyclopedia*, 2508–09.

Pecan

1. F. H. Burnette et al., *Pecans*, Louisiana State Agricultural Experiment Station Bulletin 69 (1902), 878, https://babel.hathitrust.org/cgi/pt?id=hvd.32044107232365&view=1up&seq=3.
2. H. P. Stuckey and Edwin Jackson Kyle, *Pecan-Growing* (New York: Macmillan, 1925), 12, https://babel.hathitrust.org/cgi/pt?id=uc1.$b308983&view=1up&seq=27.
3. Stuckey and Kyle, *Pecan-Growing*, 12; Jules Janick and Robert E. Paull, eds., *The Encyclopedia of Fruit & Nuts* (Cambridge, U.K.: Cambridge University Press, 2008), 421.
4. J. Russell Smith, *Tree Crops: A Permanent Agriculture* (New York: Devin-Adair, 1953), 243.
5. Burnette et al., *Pecans*, 850.
6. C. A. Reed, *The Pecan*, USDA Bulletin 251 (July 12, 1912), 16, https://www.biodiversitylibrary.org/item/83395#page/7/mode/1up; Stuckey and Kyle, *Pecan-Growing*, 1.
7. Burnette et al., *Pecans*, 878–79; Susanna Lyle, *Fruit & Nuts* (Portland: Timber Press, 2006), 110.
8. Reed, *The Pecan*, 7–8.
9. "How Pecans Are Grown," The US Pecan Growers Council website (2018), https://uspecans.org/how-pecans-are-grown.
10. Robert T. Morris, *Nut Growing* (New York: Macmillan, 1921), 174–75.
11. "How Pecans," US Pecan Growers; Lyle, *Fruit & Nuts*, 111.
12. Smith, *Tree Crops*, 255.
13. Ed Perry and G. Steven Sibbert, *Harvesting and Storing Your Home Orchard's Nut Crop*, University of California Division of Agriculture and Natural Resources Publication 8005 (1998), 6, http://homeorchard.ucdavis.edu/8005.pdf.
14. Michael Dirr, *Manual of Woody Landscape Plants*, 5th ed. (Champaign, IL: Stipes Publishing, 1998), 185.
15. Janick and Paull, *Encyclopedia of Fruit*, 423.
16. William Reid, "Pecan Pollination Season," *Northern Pecans* (blog), May 20, 2020, http://northernpecans.blogspot.com/2020/05/pecan-pollination-season.html; Janick and Paull, *Encyclopedia of Fruit*, 423.
17. Reed, *The Pecan*, 11.
18. L. H. MacDaniels, "Nut Growing in the Northeastern States," *Arnoldia* 1, no. 9–12 (October 31, 1941), 46.
19. Burnette et al., *Pecans*, 859.
20. Burnette et al., *Pecans*, 858.
21. William Reid, "Bearing Pecan Trees Compete with Ground-Cover for Water," *Northern Pecans* (blog), July 2, 2020, http://northernpecans.blogspot.com/2020/07/bearing-pecan-trees-compete-with-ground.html.
22. Reed, *The Pecan*, vii.
23. William Reid, "Stinkbugs Create a Nasty Surprise," *Northern Pecans* (blog), January 14, 2014, http://northernpecans.blogspot.com/2014/01/stinkbugs-create-nasty-suprise.html.

Red and White Currants

1. U. P. Hedrick, *Sturtevant's Notes on Edible Plants—Red Rubrum* (1919), https://uses.plantnet-project.org/en/Ribes_(Sturtevant,_1919).
2. Fred W. Card, *Bush-Fruits* (New York: Macmillan, 1914), 379.
3. U. P. Hedrick, *The Small Fruits of New York* (Albany: J. B. Lyon, 1925), 244, http://www.chathamapples.com/SmallFruitsNY/CurrantEvol.htm.
4. Hedrick, *Small Fruits*, 244.
5. Hedrick, *Sturtevant's Notes.*
6. Reich, *Uncommon Fruits*, 180.
7. Susanna Lyle, *Fruit & Nuts* (Portland: Timber Press, 2006), 381.

8. Card, *Bush-Fruits*, 340.
9. John Torgrimson, *Fruit, Berry and Nut Inventory*, 4th ed. (Decorah: Seed Savers Exchange, 2009), 251–54; Lee Reich, *Uncommon Fruits for Every Garden* (Portland: Timber Press, 2004), 184–86.

Red Raspberry

1. Fred W. Card, *Bush-Fruits* (New York: Macmillan, 1914), 191.
2. U. P. Hedrick, *The Small Fruits of New York* (Albany: J. B. Lyon, 1925), 51.
3. Hedrick, *The Small Fruits*, 49; Jules Janick and Robert E. Paull, eds., *The Encyclopedia of Fruit & Nuts* (Oxfordshire, U.K.: CAB International, 2008), 752–53.
4. Susanna Lyle, *Fruit & Nuts* (Portland: Timber Press, 2006), 399; Card, *Bush-Fruits*, 186; Janick and Paull, *The Encyclopedia of Fruit*, 751–52.
5. Hedrick, *The Small Fruits*, 86–152.
6. Lori Bushway et al., *Raspberry and Blackberry Production Guide for the Northeast, Midwest, and Eastern Canada* (Ithaca, NY: Natural Resource, Agriculture, and Engineering Service, 2008), 3; Hedrick, *The Small Fruits*, 47; Dennis W. Magee and Harry E. Ahles, *Flora of the Northeast* (Amherst: University of Massachusetts Press, 1999), 605; Janick and Paull, *The Encyclopedia of Fruit*, 749.
7. Janick and Paull, *The Encyclopedia of Fruit*, 752; Lyle, *Fruit & Nuts*, 401.
8. Jammie Favorite, *American Red Raspberry*, USDA Natural Resources Conservation Service Plant Guide (January 29, 2003), https://plants.usda.gov/DocumentLibrary/plantguide/pdf/cs_ruid.pdf; M. Grieve, *A Modern Herbal*, vol. 2 (New York: Dover Publications, reprint 1971, originally published in 1931), 671.
9. Bushway et al., *Raspberry and Blackberry*, 105–06.
10. David T. Handley, *Growing Raspberries and Blackberries*, University of Maine Cooperative Extension Bulletin 2066 (2006).
11. Bushway et al., *Raspberry and Blackberry*, 17, 92.

Schisandra

1. Susanna Lyle, *Fruit & Nuts* (Portland: Timber Press, 2006), 412–13.
2. Richard M. K. Saunders, *Monograph of Schisandra (Schisandraceae)*, vol. 58, *Systematic Botany Monographs* (Ann Arbor, MI: American Society of Plant Taxonomists, 2000), 95.
3. Saunders, *Monograph of Schisandra*, 99.
4. Dennis J. McKenna et al., *Botanical Medicines: The Desk Reference for Major Herbal Supplements* (London: Routledge, 2002), 888.
5. Jules Janick and Robert E. Paull, eds., *The Encyclopedia of Fruit & Nuts* (Oxfordshire, U.K.: CAB International, 2008), 848; Christine Leon and Lin Yu-Lin, *Chinese Medicinal Plants, Herbal Drugs and Substitutes* (Richmond, Surrey, U.K.: Kew Publishing, 2017), 740.
6. McKenna et al., *Botanical Medicines*, 887.
7. McKenna et al., *Botanical Medicines*, 888.
8. McKenna et al., *Botanical Medicines*, 887–919.
9. Leon and Yu-Lin, *Chinese Medicinal Plants*, 740–43; Lyle, *Fruit & Nuts*, 413.
10. McKenna et al., *Botanical Medicines*, 888.
11. Saunders, *Monograph of Schisandra*, 30.
12. Saunders, *Monograph of Schisandra*, 1.
13. Sonia Schloemann, "Cultural Requirements for Cultural Production of *Schisandra chinensis*, We Wei Zi or Chinese Magnolia Vine," University of Massachusetts Agricultural Extension (2009), https://ag.umass.edu/sites/ag.umass.edu/files/fact-sheets/pdf/schisandra_project_report.pdf.
14. Schloemann, "Cultural Requirements."

Seaberry

1. Frank Kingdon Ward, *The Land of the Blue Poppy* (Little Compton, RI: Theophrastus

Publishing, reprint 1973, originally published in 1913), 92.

2. Thaddeus McCamant et al., "Perennial Fruit: Seaberries," Minnesota Institute for Sustainable Agriculture (2018), 79, http://misadocuments.info/seaberries.pdf.
3. Jim Todd, "Introduction to Sea Buckthorn," Ontario Ministry of Agriculture, Food and Rural Affairs (2016), http://www.omafra.gov.on.ca/english/crops/facts/seabuckthorn.htm.
4. Floraleads GR, "Sea Buckthorn," Natural Health and Beauty Products website, accessed October 8, 2021, http://floraleads.com/seabuckthorn/; Jules Janick and Robert E. Paull, eds., *The Encyclopedia of Fruit & Nuts* (Oxfordshire, U.K.: CAB International, 2008), 339.
5. Susanna Lyle, *Fruit & Nuts* (Portland: Timber Press, 2006), 241–43; Georgetown University Medical Center, "Sea Buckthorn," accessed October 8, 2021, https://sites.google.com/a/georgetown.edu/urban-herbs/seabuckthorn.
6. Lyle, *Fruit & Nuts*, 241–43.
7. Georgetown University, "Sea Buckthorn."
8. McCamant et al., "Perennial Fruit," 80.
9. Lyle, *Fruit & Nuts*, 241–43; Alina Petre, "Top 12 Health Benefits of Sea Buckthorn Oil," Healthline Media, December 5 2018, https://www.healthline.com/nutrition/sea-buckthorn-oil; Thomas S. C. Li, "Sea Buckthorn: New Crop Opportunity," in *Perspectives on New Crops and New Uses*, edited by J. Janicky (Alexandria: ASHS Press, 1999), https://hort.purdue.edu/newcrop/proceedings1999/pdf/v4-335.pdf.
10. Michael Dirr, *Manual of Woody Landscape Plants*, 5th ed. (Champaign, IL: Stipes Publishing, 1998), 431–32; Abhay Sharma et al., "Seabuckthorn a New Approach in Ecological Restoration of Himalayan Ecosystem: A Review," *International Journal of Chemical Studies* 7, no. 1 (2019), 1221, https://www.chemijournal.com/archives/2019/vol7issue1/PartU/7-1-132-535.pdf.
11. Dirr, *Manual of Woody*, 431–32.

Shipova

1. J. D. Postman, "*Sorbopyrus auricularis* (Knoop) Schneider: An Unusual Pear Relative," *American Pomological Society Fruit Varieties Journal* 50, no. 4 (October 1996), 218–20, https://www.pubhort.org/aps/50/v50_n4_a35.htm.
2. Lee Reich, *Uncommon Fruits for Every Garden* (Portland: Timber Press, 2004), 223.
3. International Dendrology Society, "× *Sorbopyrus auricularis* (Kroop) Schneid.," *Trees and Shrubs Online* website, https://treesandshrubsonline.org/articles/x-sorbopyrus/x-sorbopyrus-auricularis/; Clemens Alexander Wimmer, "Die Bollweiler Birne × *Sorbopyrus irregularis* (Münchh.) C. A. Wimm.: Geschichte und Nomenklatur," *Zandera* 29, no. 2 (2014), 59–69, http://www.gartenbaubuecherei.de/Zandera/2014_2_Birne.pdf.
4. J. D. Postman, "Intergeneric Hybrids in *Pyrinae* (=*Maloideae*) Subtribe of *Pyreae* in the Family *Rosaceae* at USDA Genebank," *Acta Horticulturae* 918 (2011), 937–43, https://www.ishs.org/ishs-article/918_123.
5. Wimmer, "Die Bollweiler Birne," 59–69.
6. International Dendrology Society, "x *Sorbopyrus auricularis*"; C. S. Sargent, *Bulletin of Popular Information* (later *Arnoldia*) (May 12, 1925), 18; Michel Hoff, "Le Poirier de Bollwiller, *Sorbopyrus auricularis*," *Bauhinia* 20 (2007), 52.
7. "Johann Bauhin, 1541–1613, *Historia Plantorum Universalis*," Cincinnati History Library and Archives (December 22, 2006), http://library.cincymuseum.org/bot/bauhin.htm.
8. Wimmer, "Die Bollweiler Birne," 59–69; Sargent, *Bulletin of Popular Information* (1925), 18.

9. US National Plant Germplasm System, "× *Sorbopyrus auricularis*" (plant acquisition record), USDA Research Service, https://npgsweb.ars-grin.gov/gringlobal/accessiondetail?id=1558416.
10. Larry Rettig, "An Uncommon Harvest: The Shipova," Dave's Garden website, November 11, 2008, https://davesgarden.com/guides/articles/view/1795.
11. Rettig, "An Uncommon Harvest"; Wimmer, "Die Bollweiler Birne," 61.
12. International Dendrology Society, "× *Sorbopyrus auricularis*."
13. Lee Reich, *Uncommon Fruits for Every Garden* (Portland: Timber Press, 2004), 221.
14. Reich, *Uncommon Fruits*, 223.
15. John Torgrimson, *Fruit, Berry and Nut Inventory*, 4th ed. (Decorah: Seed Savers Exchange, 2009), 203.
16. Torgrimson, *Fruit, Berry*.
17. Barbara W. Ellis and Fern Marshall Bradley, eds., *The Organic Gardener's Handbook of Natural Insect and Disease Control* (Emmaus, PA: Rodale Press, 1996), 70; Fern Marshall Bradley and Barbara W. Ellis, *Rodale's All New Encyclopedia of Organic Gardening* (Emmaus, PA: Rodale Press, 1992), 257.
18. International Dendrology Society, "× *Sorbopyrus auricularis*."
19. Martin Crawford, "Rosaceae Family Intergeneric Hybrids," Agroforestry Research Trust, https://www.agroforestry.co.uk/wp-content/uploads/site-files-pdf/AGN_sample.pdf.

Spikenard

1. Allyson Levy and Scott Serrano, personal correspondence from the Hortus Arboretum and Botanical Gardens.
2. Huron Herbert Smith, "Ethnobotany of the Menomini Indians," *Bulletin of the Public Museum of the City of Milwaukee* 4, no. 1 (1923), 24, 62.
3. M. Grieve, *A Modern Herbal*, vol. 2 (New York: Dover Publications, reprint 1971, originally published in 1931), 760.
4. Smith, "Ethnobotany," 24, 62; L. H. Bailey, *The Standard Cyclopedia of Horticulture*, vol. 2, 3rd ed. (New York: Macmillan, 1930), 344.
5. Merritt Lyndon Fernald and Alfred Charles Kinsey, *Edible Wild Plants of Eastern North America* (New York: Harper & Brothers, 1958), 282.
6. Dennis W. Magee and Harry E. Ahles, *Flora of the Northeast* (Amherst: University of Massachusetts Press, 1999), 776.
7. "*Aralia racemosa*," Missouri Botanical Garden, accessed October 8, 2021, http://www.missouribotanicalgarden.org/PlantFinder/PlantFinderDetails.aspx?kempercode=v270.
8. Roger Phillips and Martyn Rix, *The Botanical Garden* (Buffalo: Firefly Books, 2002), 376.

Wintergreen

1. Euell Gibbons, *Stalking the Wild Asparagus* (Putney, VT: Alan C. Hood, 1962), 212.
2. Roger Phillips and Martyn Rix, *The Botanical Garden* (Buffalo: Firefly Books, 2002), 206.
3. Merritt Lyndon Fernald and Alfred Charles Kinsey, *Edible Wild Plants of Eastern North America* (New York: Harper & Brothers, 1958), 309; L. H. Bailey, *The Standard Cyclopedia of Horticulture*, vol. 2, 3rd ed. (New York: Macmillan, 1930), 1318.
4. Cybele May, "Life Savers Pep-O-Mint & Wint-O-Green," *Candyblog* (November 1, 2010), http://www.candyblog.net/blog/item/lifesavers_pep-o-mint_wint-o-green; Cybele May, "Classic Gums: Black Jack, Clove, Beemans & Teaberry," *Candyblog* (June 10, 2009), https://www.candyblog.net/blog/item/classic_gums.

Index

B

S

About the Authors

Emmett Serrano

Allyson Levy and **Scott Serrano** are both exhibiting visual artists and co-founders and directors of Hortus Arboretum and Botanical Gardens in New York's Hudson Valley. Their garden began as a source of inspiration and raw materials for their art. Over time their interest in growing a wider selection of plants grew until the garden encompassed 21 acres (8.5 ha) and developed as their primary passion. Along the way they began planting a vast diversity of plants, both edible and ornamental. This grew into an extensive collection of cold-hardy cactus, magnolia trees, viburnums, and fruit and nut trees, focusing on the rare, underutilized plants. The arboretum is now a nonprofit organization and Level II arboretum.

the politics and practice of sustainable living

CHELSEA GREEN PUBLISHING

Chelsea Green Publishing sees books as tools for effecting cultural change and seeks to empower citizens to participate in reclaiming our global commons and become its impassioned stewards. If you enjoyed *Cold-Hardy Fruits and Nuts*, please consider these other great books related to gardening and edible landscaping.

THE HOME-SCALE FOREST GARDEN
How to Plan, Plant, and Tend a Resilient Edible Landscape
DANI BAKER
9781645020981
Paperback

THE HOLISTIC ORCHARD
Tree Fruits and Berries the Biological Way
MICHAEL PHILLIPS
9781933392134
Paperback

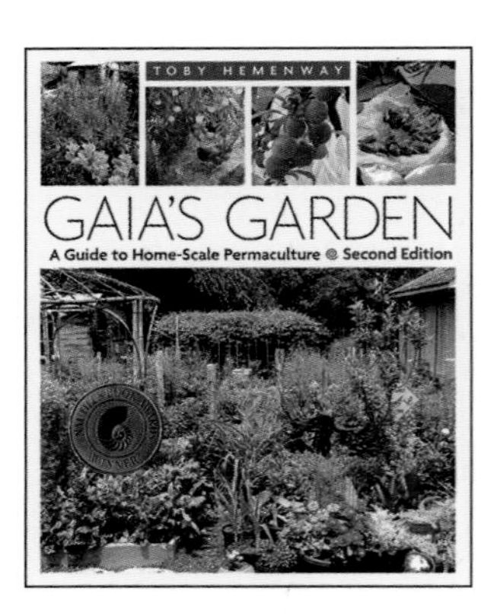

GAIA'S GARDEN
A Guide to Home-Scale Permaculture, Second Edition
TOBY HEMENWAY
9781603580298
Paperback

PERENNIAL VEGETABLES
From Artichokes to 'Zuiki' Taro, A Gardener's Guide to Over 100 Delicious, Easy-to-Grow Edibles
ERIC TOENSMEIER
9781931498401
Paperback

For more information,
visit **www.chelseagreen.com**.